T0310815

Scientific Writing 3.0

A Reader and Writer's Guide

Scientific Writing 3.0

A Reader and Writer's Guide

Jean-Luc Lebrun and Justin Lebrun
Scientific Reach

 **World Scientific**

NEW JERSEY · LONDON · SINGAPORE · BEIJING · SHANGHAI · HONG KONG · TAIPEI · CHENNAI · TOKYO

Published by

World Scientific Publishing Co. Pte. Ltd.

5 Toh Tuck Link, Singapore 596224

USA office: 27 Warren Street, Suite 401-402, Hackensack, NJ 07601

UK office: 57 Shelton Street, Covent Garden, London WC2H 9HE

British Library Cataloguing-in-Publication Data
A catalogue record for this book is available from the British Library.

SCIENTIFIC WRITING 3.0
A Reader and Writer's Guide

ISBN 978-981-122-883-4 (hardcover)
ISBN 978-981-122-953-4 (paperback)
ISBN 978-981-122-884-1 (ebook for institutions)
ISBN 978-981-122-885-8 (ebook for individuals)

For any available supplementary material, please visit
https://www.worldscientific.com/worldscibooks/10.1142/12059#t=suppl

Typeset by Stallion Press
Email: enquiries@stallionpress.com

Printed in Singapore

Preface

You are a scientist. You have spent years educating yourself and improving your research skills. Years of perfecting new technologies and methodologies, years of data collection and organization, and years of adapting to new challenges. Every hour spent in the lab, at the library, or at home in front of your computer has broadened your knowledge and capacity in your field. With dedication, persistence, and occasional failure, you have discovered something new — something unknown to the world. You have pushed the boundaries of science just a little, but you can justifiably say that you are among the world's experts in your niche field — possibly the sole expert. But what good is it to be the world's sole expert? Alone, you are ill-equipped to shoulder the burden of scientific progress. So you uncap your pen (or pull out your keyboard) and start writing a letter (your paper) to other scientists.

Herein lies a problem. Writing — the task in front of you — is not research-based. You may feel intimidated. After all, you probably have not received specific training to face the many challenges of peer-to-peer communication through academic journals.

At this juncture, you have two options. The first path is the road most travelled: You do not consider you lack expertise in writing. After all, have you not already written hundreds of thousands of words over the course of your education? Have you not already published several theses at the bachelor, master, and Ph.D. level? And if you do lack expertise, do you not have an ocean of examples in front of you in hundreds of academic journals, papers that you can study and learn from or emulate? Or does it even matter whether you are a great writer? You chose to do science, not literature.

The second path is the road less traveled: You want to gain expertise in writing, become a better communicator to enhance your publication chances. It is the path that you, by picking up this book, are considering.

Would it surprise you if we told you that both paths lead to the same destination? One road winds back and forth, passing through metaphorical mists, moats, and brambles. It takes years to travel, and is littered with paper rejections and writing hardship. At its end, you may have mastered many aspects of writing, but at what personal cost? The other path is much shorter. It is a steep uphill climb for a few weeks, but from its summit, you can clearly see the brambles, the moats, and the mists... and walk right past them.[1]

This book is your roadmap to the top of the hill. Reading it and practicing the recommended techniques are the steep uphill climb. You will need to flex your intellectual muscles, not walk the talk, but walk the thought, put it in motion through our specially crafted exercises. You may even feel sore for a while. But that path is worth following.

You can choose to master many skills during your lifetime. Some will be helpful in highly specific situations (replacing a car tire), while others will find use in broader situations (swimming). But certain skills affect so many facets of your life that it would be almost negligent not to master them. Writing is one of them.

The book you are holding today is not the same roadmap we initially presented in the 2007 or 2011 editions. Today's book is Scientific Writing 3.0 — the third chronicle of our constantly evolving research into the minds of readers and writers. It has been updated with new examples, new stories, and new discoveries. While the initial versions of the book focused on writing techniques, this new edition also discusses the writer's attitude towards the audience, strategies for getting published, objective measures of readability, and the predominant scientific writing style, with its characteristics and pitfalls.

You're on the right path — now take the first step.

---

[1] This isn't a hypothetical encouragement- we've heard this exact feedback from senior scientists who have expressed the regret of not having read this book earlier in their careers.

Contents

Part 1

The Reading Toolkit

This title probably conjures up the image of a schoolboy's pencil-case containing a few chosen items designed to help reading: a pair of glasses for better readability of the footnotes and formulas, a bookmark to help with our failing memory, a bag of pekoe tea to boost our attention when it declines, a LED flashlight to continue reading in the dark when away from electricity. The reading toolkit I have in mind, however, is filled with resources of the invisible kind: time, memory, energy, attention, and motivation. A skillful writer uses this toolkit to minimize the time, memory, and energy needed for reading while keeping reader attention and motivation high.

Chapter 1

Writer vs. Reader, a Matter of Attitude

Allow me to begin this book by asking you a rather startling question: who are you? A reader or a writer?

You're *reading* this book, so you must be a reader.

And you're reading *this book* so you must be a writer.

More precisely, you are both a reader and a writer scientist. You read papers published in academic journals to deepen your knowledge of the field, learn about recent advances or breakthroughs, or even borrow certain steps from an existing protocol. You may occasionally read out of curiosity, but you never read academic papers for fun. You are a reader with a purpose, a need to fulfil. As for being a writer scientist, you don't write papers for fun either. So why do you write?

Allow me to introduce you to the archetypal science writer, John.

John enjoys research. He would love for his work to be recognized and to lead his own team as a PI someday. More pressingly, his boss requires him to hit certain key performance indicators (KPI), including getting a number of papers published in high-tier journals each year. So John prepares his papers with the following idea in mind: "I've done the research. Now I need to write down what I've done so that my work can be documented, officially recognized through citations, and help me hit my KPI."

Although John's reasoning may seem logical at surface level, a deeper look reveals several serious issues. Stop reading for a minute, and try to identify the key problem with John's mindset.

...

All done? Hopefully you've identified the main culprit: John focuses only on himself and his needs. He wants his work to be documented. He wants to be recognized and to hit his KPI. To accomplish these self-interested goals, he will write a paper. His goal is **writer-centered, not reader-centered, and this attitude introduces tangible flaws in his writing.**

Why Am I Writing This Paper?

Before we entertain further thought into the *what* and the *how* of John's flaws, we need to zoom all the way out and answer the fundamental questions that guide his pen.

As a researcher, John must write scientific papers. But why?

Because he needs to hit his KPI targets to keep his job. **But again, why?**

Because having his work published in journals puts it within reach of other scientists who will use it to further their own research. In doing so, he can indisputably show that he has created value for the scientific community.

And that, ultimately, should be the reason that John is a researcher. He did not take on a career in research because he enjoys the challenges of hitting KPIs — he took on a career in research because of a fundamental interest in furthering science, of making a unique contribution to humankind's existing knowledge.

The writer-centered approach has John at the base of the pyramid, looking only to reach the next rung. It takes a myopic view of the diagram below, confusing the objective for the end goal.

The longer you spend in research, the easier it becomes to focus on the incremental next step and forget the big picture. **Do not**

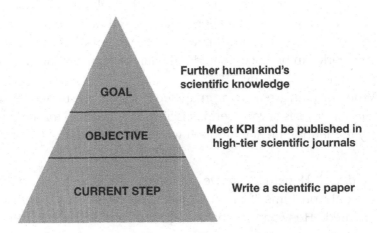

write to survive academia. Write to be useful to others. Although seemingly unimportant, this little change in attitude nevertheless could protect against the tangible writing flaws we next describe.

The Illusion of Clarity

If you're an expert on your topic, writing should be easy, shouldn't it? After all, you have the knowledge, so all you need to do is "write what you think".

But is it really that simple? Read the following paragraph, inspired from an abstract[1]:

Changes in the gut microbiota have been known to have modulatory effects on host metabolisms. These changes have been demonstrated in GEMM, but never in wild type. The current study aims to investigate whether non-congenic strains reflect similar findings to initial studies.

Was that paragraph clear to you? Your answer could be "yes" or "no". It depends on you, the reader, and your level of familiarity with the field and its vocabulary. While the passage is likely *clear* to most biologists, it may be less *clear* to other scholars, for example those studying hydraulics. The word "clear" is ambiguous. It is an adjective, and like most adjectives, it is subjective. What is clear to you may not be clear to me, and vice-versa.

---

[1] Don't take this paragraph to heart! The science is nonsensical and for illustrative purposes only.

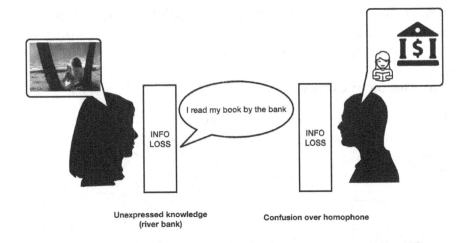

**Unexpressed knowledge
(river bank)**

Confusion over homophone

Do not assume that your writing will be clear to the reader simply because it is clear to you.

In fact, with the passage of time, authors can even be unclear to themselves!

It's Friday evening and you have just finished writing a rather challenging technical paragraph. You think to yourself, "It took some work, but I'm glad I finally managed to express my ideas clearly!" But when you return to your desk on Monday and re-read that "clear" paragraph, you find it to be extremely difficult to understand. What happened?

Since the text itself did not change over the weekend, the change must have occurred in you. On Friday evening, *you read as a writer*: any required knowledge or background needed to understand the text remained fresh in your mind, filling in the gaps in understanding. But the passage of time and disengagement with the content emptied that buffer over the next few days, so that on Monday morning, *you read as a reader*.

Thinking reader[2] when you write makes your writing clearer and more accessible — two qualities reviewers and editors greatly value. Thinking reader is not as selfless as one might think. By

[2] Lebrun, Jean-luc. *Think Reader — Writing by Design: Reader-based techniques to improve your writing.* Scientific Reach, 2019 https://www.amazon.com/THINK-READER-Writing-Reader-based-techniques/dp/173389750X

helping the reader understand your paper, you save the reader's time, avoid reader frustration, and increase reader satisfaction, all of which increase the probability that the reader uses, shares, and cites your work.

The Do-unto-others-as-others-do-unto-you Inversion

Ironically, we find reading most scientific texts somewhat unpleasant. But when comes our turn to write one, we espouse the same unpleasant style of writing. We dislike overabundant data but include it in our own papers. We dislike having to hunt through documents for the meanings of acronyms, but are not shy to use acronyms liberally in our own writing. Why must we act like the writers whose work we dislike reading?

Some scientists use dense prose and intimidating vocabulary, assuming that such writing conveys the level of expertise and sophistication necessary to be published. "Complexity is inevitable and scientific," they say. "If the complexity of science makes it difficult to communicate, so be it."

These scientists, however, are often proven wrong. There are wonderful examples of Nobel Laureate scientists[3] describing the most complex of topics in the most comprehensible ways. Complexity creates a reader-writer chasm that often translates into inaccessibility. The chasm calls for a bridge; Alas, writers all too often rely on their readers to build it.

Though these words may sting, they are difficult to deny. In the absence of formal training on how to write a scientific manuscript, many turn towards published journal papers for examples of good writing. Learning by example is a worthy pursuit, but how well can one learn when the source of examples is itself plagued by many poorly written papers? Should we not try to raise the standards of scientific writing? After all, you did not pick up this book to be an average scientific communicator; you picked it up to be an exemplary one. The golden rule applies: **Do unto others as you would have them do unto you.** Reader-centered is the right attitude.

---

[3] See R. Hoffmann, *The Same, Not The Same,* Columbia University press, 1995

The Right(er) Attitude

You now know to avoid a writer-centered attitude, so let's address what it means to adopt a reader-centered mindset. Unfortunately, you will rarely have the opportunity to look your readers in the eye and tell them that you care about them and their reading experience. At best, you can communicate your care for them through clear, purposeful writing. What is this purpose you need to achieve? Just sign your name on the line below to find out.

Hi reader, it's me, _____. I want you to directly benefit from my work, carry it forward, and be successful. I want to inspire you, be the spark that ignites new research ideas, opens new research avenues.

I am aware that reading is accessory and that your time is precious. I wish to save you time by structuring my writing such that you can rapidly find what you need. I will be concise. I will write in a way that allows the sentences to flow into your brain like a wide river flowing into the sea — without turbulence, cataracts, boulders, or deep meanders. I will not waste your time by concocting misleading titles and promises that I cannot deliver. I want to give you enough details for you to reproduce my work. I do not wish to mislead you, and will therefore not cook up numbers, doctor images, or sift p-values.

Likewise, I will do everything I can to make my writing clear, by anticipating the gaps in your knowledge and avoiding obscurity in expression. I will do my utmost to travel alongside you, not lose you, to guide you where I foresee learning difficulties. I want to write with a clarity that not only allows you to think "I understand", but expressly forbids any misunderstanding or ambiguity. That is the level of clarity I wish to reach, even though I know I will fail here and there.

Dear reader, I am grateful that you chose to read my paper when so many other papers are awaiting your attention. And since I have your attention, I will not waste it by distracting you, by unduly calling it on trivial matters. Rather, I will guide it directly to what really matters and should be of most interest to you.

I know that it is by helping you achieve your research goals that my own research will be justified through your citation. In short, dear reader, I do not exist without you. I will trace a path forward, but it is only when you walk upon it that it has value.

Chapter 2

Strategic Writing

Height, skill, agility — these are the necessary attributes of a *good* basketball player. But to be a *great* basketball player requires much more than skills and being a stellar physical specimen — it requires a mastery of the rules and plays of the game as well as an understanding of how the opposing team mobilizes. In other words, it requires strategy.

You may be a naturally talented writer. But publishing research is not simply a matter of writing well. You need to understand that your manuscript is a product and that this product must be attractive, marketable, and on the right shelves. You need to understand the needs and wants of your consumers/readers and how to fulfil them. You need to build trust so that your readers not only like your writing but also find your work credible. And you certainly need to consider the most demanding of your readers — the editors and reviewers.

The Scientific Paper: An Intellectual Product

Your scientific paper is a product. It took time and effort to produce, and it will be consumed by its readers through the act of reading. Whereas consumed food yields energy, consumed papers yield knowledge. But just as getting a new food product into the mouths of consumers is challenging, so is getting a new intellectual product into the minds of learners.

The table below tells two stories: one is about a food product, and the other about a scientific paper. If you wish to read the two stories sequentially, read from top-to-bottom in the left column first, then top-to-bottom in the right column. If you wish to read both stories in parallel, read from left to right.

Imagine you are the inventor of a new kind of breakfast cereal, which is different from the other types on the market. Instead of a flake or a loop, each piece is either a number between 1 and 9 or a mathematical symbol: +, −, ×, or /. Your hope is that this cereal will encourage children to practice simple mathematics with each scoop of the spoon into the bowl.

Imagine you are the writer of a new scientific paper, which contains an innovative contribution to the existing scientific literature.

While you enjoy eating your own cereal, you would much rather have it consumed by others as well. You realize that in order for this product to be sold, you need to get it on the shelves of stores.

While you're proud of the pdf sitting on your desktop, you would much rather have it read by other scientists. You recognize that in order for this paper to be downloaded, you absolutely need to get it into an upcoming issue of an academic journal.

Which store will you choose? Will you try to get your product on the shelves of a major distributor like Carrefour or Whole Foods? Or should you aim to get your product carried by a smaller but more specialized store with a knowledgeable clientele who will appreciate the cleverness of your product?

Which journal will you choose? Will you try to get your paper published in a major journal like Science or Nature? Or should you aim to be published in a smaller but more specialized journal whose readers you know will be directly interested in your findings?

How can you convince a store to pick up and stock your product on their shelves? You have to convince them that it will benefit them and their customers (valued and purchased by their customers, and therefore profitable for them).

How can you convince a journal to publish your paper? You have to convince its editors that it will benefit them and their readers (valued and cited by their readers, and therefore increasing the journal's impact factor — and indirectly profits).

What form does this convincing take? You have to meet up with the store owner to show them your product and argue your case for why you believe it is valuable and their customer base would buy it.

What form does this convincing take? You have to submit your manuscript to the editor, and argue in a cover letter why you believe the readers of this journal would be interested by your discovery.

But you're not just going to take a handful of cereal out of your pocket to show the owner, are you? You're going to present a package- a nicely designed cardboard box. On one side of the box is the title of your product, and on the back is some writing that succinctly describes the value and uniqueness of the mathematical cereal inside.

But you're not going to submit just the manuscript itself for review, are you? You have to present it with a title and abstract that quickly and accurately encapsulates the value and uniqueness of the results inside.

Once the store owner has your well-designed package in their hands, will they rush to put it on their shelves? Not yet! While they may know enough to see the value from the package, they don't have the expertise to determine if the product is safe for consumption. Before they make any kind of shelf stocking decision, they need to send the product for quality control and for taste testing.

Once the editor has your title, abstract, and paper in their hands, will they rush to publish it in their journals? Not yet! While they may know enough to see the value from the cover letter, title, and abstract, they don't always have the expertise to determine if the methodology is sound or the discussion robust. Before they make any kind of publishing decision, they need to send the paper for peer review.

If the experts deem the cereal worthy, the store owner may finally decide whether or not to stock it on their shelves.

If the reviewers deem the paper worthy, the editor will finally consider whether or not to publish it.

Will the store advertise your product for you? Yes, for a short while. They may add stickers or displays that otherwise emphasize that the product is new, but these will disappear after a few weeks. Once this period is over, your cereals will no longer be actively promoted.

Will the journal advertise your paper for you? Yes, for a short while. They may give free access to your paper for a limited time to a limited number of people of your choice, or they may feature it in an editorial (it picks up traction and affects their impact factor[1]). Once this period is over, your paper is no longer actively promoted.

When on the shelves, does your product get preferential treatment? No, it simply joins the rest in a wall of cereals that the consumer walks past when browsing for something to buy. It is therefore very important that the packaging makes the product stand out and catches the shopper's attention. The shopper doesn't need to pay to read the packaging, just to eat the product.

Once published, does your paper get preferential treatment? No, it simply joins the list of other paper titles that are drawn up when a user queries a search engine like google scholar on a keyword related to your work. It is therefore very important that the title and abstract stand out and catch the reader's attention. The reader doesn't have to pay the journal to read the title and abstract, just to download the full paper.

As we have demonstrated, many parallels exist between physical and intellectual products. To be a successful author, it is essential that you grow to understand not just the process of writing a paper, but the entire ecosystem within which that writing is promoted and consumed. But it isn't enough just to understand—you must also know

---

[1] The impact factor of journals is calculated over 3 years from the date of publication. After 3 years, the impact factor remains fixed even if citations continue to grow.

how to use this understanding to your advantage, as we will illustrate throughout the rest of this chapter.

Product expiration date

If you wish to avoid intestinal trauma, checking the expiration dates on food products is always a good idea. But since intellectual products are consumed by the mind and not the gut, is it possible for intellectual products to expire? Can you experience a case of mental trauma?

Intellectual products expire much in the same way that paracetamol expires. Not immediately dangerous, but less and less effective, to the point where it no longer helps. When first published, your scientific contribution will be truly novel, and open the mind of your readers to new possibilities. As time passes, the novelty wears off as it is incorporated into practice or future research, to the point where there is no longer much of a need to refer to the original paper anymore. So perhaps we shouldn't say that the paper has an *expiration* date, but a *best before* date. How long is this period for a scientific paper? It depends on how quickly the field moves forward[2], but generally a few years[3].

Journals are interested in promoting your papers, but they treat the *best before* date as a deadline. In the first three years, any journal promotion helps you, but also helps them with their impact factor. After that, the promotion is over. But should a paper no longer be promoted simply because it no longer helps the journal? After all, papers are written for readers, not journals, and it would be ridiculous to assume that a paper loses its value to *all* readers after three years. This is what a citation growth chart might look like for an article from a particular writer:

---

[2] E.g. slower for physics, faster for biochemistry.
[3] Different types of papers also have different best before dates. Review papers, for example, tend to have longer shelf lives as their purpose is not only to highlight novelty but also to act as a go-to resource for incoming researchers in a field.

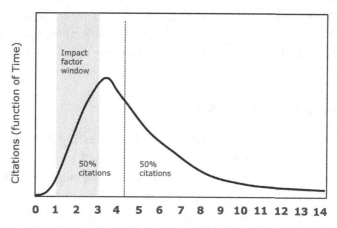

Number of years after publication

As you can see, there is an initial bump from the novelty effect and journal promotion. Afterwards, the number of new citations decreases slowly in what is called *the long tail*. Gradually accumulating citations here-and-there for your older papers isn't as exciting as getting cited on your recent work, but these citations are no less valuable. You, personally, do not have a 3-year impact factor window to consider. Every citation, new or old, is a currency that paves your way to grants and tenure. And think of the readers who stumble upon your older work for the first time, and see real value in it. To them, it isn't an old paper — it is a new paper, new knowledge that allows them to validate their ideas or justify their approach. How can you encourage these readers? Are you limited to hoping that they will stumble upon your paper through a keyword search a few years down the line?

As the producer of your intellectual product, you have to also take responsibility for its dissemination. Leaving it in the hands of the journals is all fine and well, but they do not ultimately care about your research. No one will ever be as passionate or as effective an advocate of your science as you. You may not have the reach or the readership of the journals, but you have many more directly relevant connections with people in your field, including people whose work you've referenced or whose papers you've read. Do not waste your social reach.

If you cite a paper, chances are, the writers of that paper will look at your paper (I do). If they find things that they can use in their future papers, your work could then be cited. This is the most direct path to future citations, therefore be highly selective in the papers you cite — in other words, who you want to be cited by. A citation is a reward, and rewards don't go unnoticed.

Promoting with social media?

While traveling in Australia, I picked up an interesting anecdote about the power that scientists could wield if they used social media effectively to publicize their work. A researcher at the Australian National University was curious enough to measure how effective tweeting about her paper would be in terms of garnering citations. It was an older paper outside the 3-year impact factor window. If you were to place it on the chart above, it would be on the declining curve after the peak. She tweeted about it, and was pleasantly surprised to find that the number of citations increased significantly — back to the previous peak. That is the power of one tweet for one paper — **a substantial increase of its citation count**.

Now of course, if you take to your long-inactive twitter account and start tweeting about old papers, you cannot expect a similar result. Her tweet was impactful because she had invested time and energy into building up a network of peers and potentially interested readers in her social media following. But considering that such a network yields repeat returns for every subsequent paper published, I would argue that it is well worth the effort.

Other social media options exist if twitter isn't your cup of tea (although if your goal is citation maximization, you should be on all major platforms). Researchgate[4] is the social media of science. You can publicize your papers there, connect with other scientists, join discussions with the authors of papers you've read, and even get statistics on who's reading or requesting to read your work. It takes very little time to create or maintain an account, and the 21st century scientist has no good reason not to be there. And while you are at it, create a profile on Academia[5]. I find it as useful as Researchgate, and upload my papers to both platforms.

---

[4] www.researchgate.net
[5] www.academia.edu

Depending on your topic and its potential popular interest, you may also be interested in being active on reddit communities such as /r/science. Do note it is a much more casual network than Researchgate, so your writing style must adapt to the medium. Reddit is where scientists and science enthusiasts go to hang out and share their stories — not just talk business! Leave your academic English at the door as much as possible, and do not consider yourself a writer as much as a conversationalist. If you're excited about a recent discovery you've made and you think people might be curious as to its applications or want details, feel free to take it to the next step and hold your own AMA (Ask Me Anything).

Build a network and use it. You are the best advocate of your science, and taking an active role in its promotion will benefit you personally, but also science as a whole.

Promoting by archiving your paper on preprint servers/repositories

You could also archive a version of your paper to a preprint server prior to peer-review or publication.[6] It sets up anteriority, and reduces the chances to be scooped by others, especially when peer-review stretches over a long period which could exceed a year. It also makes your paper visible enough to gather early feedback. The practice is usual in the physical sciences (ArXiv) and spreading to new domains (BioRxiv, Chemxriv, PsyArXiv).

Journal Choice: Subscription or Open?

You have probably heard of the open access movement in scientific publishing. There is already a wealth of information on its history, models and motivations available on the internet, so we won't unnecessarily paraphrase what can easily be accessed from a Wikipedia article. Instead, we'll focus only on the essential details and what concerns you as a writer seeking to be published.

---

[6] https://doi.org/10.3390/publications7020034 "Ten hot topics around scholarly publishing" (2019)

The key differences between a subscription-based journal (the "traditional journal") and an open access journal are the issues around payment, copyright, and prestige. Let's take a look at payment and copyright first.

Payment and copyright

In traditional journals, the publisher pays all costs associated with the publication: layout, editing, printing, distribution, etc. In exchange, the author hands over the copyright to the journal, and the paper is placed behind a paywall. To access the article, potential readers have to pay a fee (either per article or as a yearly subscription). **Publishing in traditional journals is free for the author, but costly to readers and academic institutions.**

In the open access publishing model, the author pays a fee to keep the paper free for readers. Fees range anywhere from USD $8 to $3900. The author keeps the copyright to the published material, and the article is free to access. **Publishing in open-access journals is free for readers but costly for the author.**

As supporters of the open science movement, we would argue that the point of publishing should be reader-focused. Not all readers (or even institutes!) are capable of shouldering the growing cost of maintaining multiple journal subscriptions just to stay abreast of the latest developments in their fields of research. Additionally, publishing costs are rarely paid for by the author out-of-pocket (a little over 10%)[7]. The employer or funding agency usually shoulders the burden. Some open access journals will even reduce or remove the fee altogether if the researcher is unable to pay or is from a developing country.

Let's think about the issue mathematically. Let's say it costs a non-open access subscription-based journal $900 to publish a paper. Let's also assume that each reader needs to pay $35 to download that paper. What number of readers does the publisher need to break even?

---

[7] Dallmeier-Tiessen, S. *et al.* Preprint available at http://arxiv.org/abs/1101.5260 (2011).

$900/$35 = 25.7, so 26 readers. At 50 readers, the journal has profited $850[8]. At 100 readers, $2,600. As the number of readers increases, the total financial cost of accessing the paper grows **multiplicatively**.

Who pays for these downloads? Researchers and research institutes, i.e. the scientific community. Where do the profits from these sales go? Rarely back into research, but to the publisher. As the number of readers increases, **the financial burden on the scientific community increases**.

By contrast, in the open access model, the financial burden to the scientific community does not scale up with the number of readers but is fixed. Although the author will pay $900 upfront and never recoup this money, their sacrifice will reach a break-even point with subscription journals at 26 readers. At 50 readers, the cost-per-access effectively drops to $18 per person. At 100 readers, $9. As the number of readers increases, the initial cost remains fixed so **the financial burden on the scientific community decreases.**

Using statistics from the highly-regarded open access journal PLoS ONE, the average article receives 800 views per year. Using a 3-year period for measurement, this means an article will be viewed 2400 times on average. In a traditional non-open access journal, this amounts to a total expenditure of 2400 × $35 = $84k for access to the research, the profits of which go to the publisher. In an open access journal, the collective cost to science remains at a fixed $900, and the paper's value is multiplied by its number of readers, so the science effectively costs $900/2400 = 37 cents per reader.

Of course, these calculations are simplified for the purpose of comparison and do not take into account other expenses, nor the fact that readership at subscription-based journals is smaller (a consideration of its own), nor that many researchers have discounted per-person fees as their employer pays a large institutional subscription[9]. But the fact remains that the amount earned by subscription

---

[8] 52 readers x $35, minus publishing cost of $900.

[9] The University of California's 10 campuses downloaded articles "nearly 1 million times" from Elsevier in 2018, and paid USD$10.5 million for the service. Even discounted from $35 per article, we can roughly work out that the discount comes to >$10 per download, a far higher number than the aforementioned 37 cents.

journals is shockingly large. Make no mistake, the world of scientific publishing is a massive for-profit industry (to put its earnings in context, it earns more than the entire recording industry[10] (in 2011, $19B > 14.8B)).

Visibility vs. Citations

Should you care more about being read or about being cited?

A 2008 (and later reconfirmed 2010) randomised controlled trial[11] found that during the first year after publication, open access articles received 119% more full-text downloads, 61% more PDF downloads, and 32% more unique visitors (i.e. not repeat visitors). Yet, the same study also found that citation count between open access and subscription-based journals was not significantly different. What does this mean? In a response to the article, Cornell researcher Philip M. Davis suggests that "The real beneficiaries of open access publishing may not be the research community, but communities of practice that consume, but rarely contribute to the corpus of literature."

This statement deserves attention. As we have expressed earlier, citations are a writer's currency. While many people may appreciate, learn from, or even use the findings in a scientific paper, only those that then go on to write about such use in further academic documents can offer a citation. Do these writers represent the totality of a journal's readers? Not every reader is a researcher for whom "Publish or Perish" is a mantra! Citations are only a visible means of having contributed to scientific academia — not the world at large.

Prestige

In an ideal world, your worth to the scientific community is measured by the worth of your scientific contribution — an entity which is extremely difficult to quantify. Especially by administrators who may not have the necessary scientific knowledge! So instead, they rely on objective heuristics that are easily measurable, such as the

---

[10] https://www.theguardian.com/science/2017/jun/27/profitable-business-scientific-publishing-bad-for-science
[11] https://doi.org/10.1136/bmj.a568

number of articles written or the ranking of the journal in which the articles were published. Traditional journals are generally more trusted due to their reputation and the long history of their publishers, such as Wiley (>200 years), Taylor and Francis (>160 years), Elsevier (>130 years), and Springer (>100 years).

In contrast, open access publishing is very young. It launched in 2002 in Budapest, gathering momentum after some government bodies demanded that all publications benefiting from Public funding be made freely available (not behind a paywall).

But does open access's youth engender a lack of trust in its model? Evidence points to a resounding "no". Open access journals are gaining prestige of their own, eliminating the benefits of the traditional model. Even the incumbent giants of the publishing industry are not immune to the open model. Just recently (March 2019), the entire University of California (UC) system[12] terminated its contract with Elsevier over rising subscription costs and limited affordable open access options. We truly are living in historic times.

Not to be left behind, publishing giants are realizing that if they do not adapt, they will lose considerable business. As a result, many large publishers have launched their own open access journals or adopted a hybrid model. The hybrid model gives the author the option of publishing traditionally or paying a publishing fee. The benefit of such a model is that the author retains the prestige of publishing in a well-ranked journal while also making their article accessible to a larger audience. The downside? Open access fees for these major publishers still tend to be on the steep side: US $3000 on average. Some journals also impose an embargo period (usually 12 months) during which the author cannot post the article on a non-commercial platform (their own webpage, for example).

In recent years, the metrics by which the prestige of a journal is measured have begun to change. Open access journals such as PLoS ONE and PeerJ are no longer simply relying on citation count to measure the success of a paper but also on social media visibility and number of readers. This shift may eventually allow scientists

---

[12] Including prestigious UC schools like UC Berkeley and UCLA (ranked 27th and 32nd respectively in QS world university rankings). UC is responsible for 10% of all US publishing output.

to publish where they believe their work will be most beneficial to readers, without having to keep career considerations and traditional measures of success in mind.

Predatory Open Access Journals

The fear of not being published is always present, but is especially strong in new investigators. So how wonderful would you feel if the following happened to you?

You open up your inbox to find an email from the XYZ Society Journal, inviting you to submit your next paper. You look up the journal online, and it seems like it has a pretty reputable board of directors. The journal has only been around for a year, but it seems like it might be up-and-coming. You decide to submit your article. A few days later, you start to worry. You just received an email from the editor explaining that your manuscript was in excellent form, and aside from a little grammatical tweak here and there, the paper is ready for publication. You still haven't received reviewer comments, nor does it seem you will (or you did, and all the reviewers wrote, "no concerns!") You always thought you were a good writer, but none of your peers had any concerns about your paper? Soon after, the editor emails you again to tell you that your paper has been accepted for publication but that you need to pay a $1500 publishing fee. Should you? Let's say you take the risk and go ahead. Your paper garners some citations over the next two years, but then one day, you receive an email from a colleague: your paper is no longer available online. Where did it go? You go to the journal's website, and it too has disappeared. Meanwhile, somewhere else in the world, a new journal has just been launched, the journal of XYZ International …

A subscription-based journal must maintain a relatively high standard of publishing and peer review, as its revenue is contingent on this. In the open access model, money comes from the authors upfront, so there isn't necessarily a strong requirement for quality down the line. Traditional journals are ranked based on their impact factor, but since this is only measured after 3 years, some predatory journals disappear once the metric comes into play, only to reopen somewhere else under a different name. And as for the "fact" that the journal's editorial board is made up of reputable members? These people may have no idea whatsoever that they are allegedly sitting on these committees!

There are many trustworthy open-access journals, including recently established ones! An open access journal that has at its heart a true commitment to furthering scientific progress is a worthy tool. But like most tools, the hand and motivation of the wielder can be destructive. It is not our intention to dissuade you from embracing open access publishing but rather to encourage careful consideration before accepting tempting offers. To combat the problem of predatory journals, new resources have emerged: blacklists and whitelists.

The first blacklist was created and updated by academic librarian and researcher Jeffrey Beall from the University of Colorado in 2008. The (very long) list included the names of journals and publishers that he considered predatory, based on his own research and the feedback of researchers who had been conned. Though Beall couldn't always distinguish between predatory and amateurish journals, resulting in some miscategorizations, the list was largely considered accurate. Unfortunately, Beall became the victim of smear campaigns directed by predatory journals, so he stopped updating the list in January 2017. His supporters have continued his work by posting updated anonymous lists on the internet.[13]

The second approach to dealing with predatory journals is the opposite of the blacklist. Most recently, associations like the Directory of Open Access Journals (DOAJ.org) and Open Access Scholarly Publishers Association (OASPA.org) have established whitelists of journals that meet their quality criteria. In this way, authors can quickly look up any journal they find questionable and see if it has already been vetted by a larger body.

So what steps can you take to protect yourself?

1. If the journal is well-known or has been around for a while, there is no need to worry.
2. If in doubt, check the whitelists to see if it has been vetted by an association such as the Directory of Open Access Journals.
3. If you don't find the journal on these lists, check the blacklist to see if it has ever been identified as predatory.

---

[13] E.g. https://beallslist.net

4. If the journal is not on any of these lists, you are likely dealing with a new journal. That does not necessarily imply it is a bad place to publish, but you will have to do a little more work first. You could read the recent issues of the journal to assess the quality of the articles published there, or you could research the editorial board and contact some of its members to inquire about their experiences with the journal or publisher. No matter what, the adage holds: if it sounds too good to be true, it probably is.

The Publishing Process

Once you have submitted your manuscript for publication, you may be tempted to rest on your laurels. Your task is finished, after all, and the manuscript is in the mail, about to be processed by the great but fairly unknown system that is the publication process. You know your paper will be handled by an editor, seen by reviewers and then either accepted for publication or not ... but not much else. How much do you trust this unknown process? How is your paper going to be evaluated? What do reviewers consider when determining the quality of your work? What makes a paper attractive to editors, and conversely, what immediately raises red flags in their minds? If you don't know the answers to these questions, how can you be confident that you have written a "publishable" paper?

Wouldn't you want to first know how (perfect) reviewers evaluate your paper?[14] Submitting it to a journal without understanding how they will evaluate it is taking a gamble. Wouldn't you rather know the rules of the game before you decide to play a hand?

You may feel crushed if you hear from your dream journal that reviewers found your contribution insufficient for publication. But instead you should feel proud that you received any feedback, negative or not. That you were reviewed indicates that you have overcome the first tier of rejections. The exact percentage depends on the journal, but a large number of papers that are submitted are never reviewed at all; they are rejected outright in the first two rounds of cuts. Surprised? After all, shouldn't all submissions be reviewed? They should not.

---

[14] Wallace, Jasmine. How to Be A Good Peer Reviewer. The Scholarly Kitchen, 2019 https://scholarlykitchen.sspnet.org/2019/09/17/how-to-be-a-good-peer-reviewer/

You've been *unsubmitted*. The first cut is made by administrative staff or an editorial assistant. You will know this fate has befallen your paper if you receive an email very soon after submission (generally within a week) telling you that your paper has been "unsubmitted". Papers are unsubmitted for a wide variety of reasons. Perhaps you did not keep to the allocated word count or forgot to include the requested author statement. Perhaps you followed the Chicago style when you were supposed to format your citations using the MLA style. For some reason or other, your paper was deemed faulty at the *submission level*, not the scientific level. What can you do? Simply follow the instructions in the journal's email reply, fix whatever mistakes might have been made, and resubmit. If the instructions are unclear, feel free to write back to the journal for clarifications; it is better to seek clarification than to waste the time of the journal with another faulty submission. The fact that your paper was unsubmitted should have little impact on your chances of having it published. At larger journals, the editor may never even know that your paper was unsubmitted in the first place.

You've been *desk rejected*. The second cut is made by an editor or a board of associate editors who deemed your paper unworthy of being sent to reviewers. The percentage of papers that fall into this category is surprisingly large (50 % to 80 % by our estimates for high impact factor journals). Not only are desk rejections now a common occurrence, they are an *increasingly* common occurrence as the number of articles submitted worldwide increases. There are only so many top-tier journals that the entire academic research pool can aim for. Occasionally, the reason for rejection will be that the level of English is too poor to allow for a fair evaluation of the science. But generally, the majority of rejections fall within two categories: inappropriate submission and or insufficiently significant submission.

Inappropriate submission: the submitted paper does not fit within the scope of the journal. Are you submitting basic research to a journal that publishes applied research? Does the journal have a history of publishing papers in your subject area, or are you trying your luck because your work is related to the kind of research that the journal publishes? Is your paper valuable only to a subset of experts in the field and not the broad readership of the journal? Is your work too specialized or too basic? Many editors complain about the volume of inappropriate submissions they receive, despite their efforts to make the journal's vision and scope transparent. Do not add to the problem by trying the shotgun approach to publishing: sending your paper to enough places, hoping one will stick. Snipe instead—the collective body of scientific editors in the world will thank you for it.

Insufficiently significant submission: the research is not significant enough to attract the attention of top tier journals. Many authors misunderstand what "significance" means here. They confuse *impressive result* with *significant result*. Let's imagine a hypothetical situation in which you're working on a voice recognition technology for people with a very heavy accent.

An increase in voice recognition accuracy from 40% to 70% would be an impressive result. But it isn't necessarily a highly significant one. In terms of usefulness, an increase in voice recognition from 90% to 97% would be a less impressive result, but a much more significant one. Both of these results would be highly valued by the scientific community, but should be published in different journals. The first result (40 → 70%) would be significant only to other researchers in the field of voice recognition and should therefore be published in a journal that caters to those specific readers. The second result (90 → 97%), pushes the technology over a tipping point of real-world usability, and is thus highly significant to a larger number of readers, including potentially those outside academia (e.g. software engineers who can now build this technology into their programs).

You can still garner many citations if you are published outside of the usual top-tier journals. Don't always aim for the top (where you are highly likely to miss), instead, aim for the appropriate tier based on significance. Desk rejections can and should be avoided by being a more discerning and careful submitter, not one who tries their luck at a game that we call the "paper submission staircase."[15]

---

[15] Have a paper to publish? Good! Try *Science* and *Nature* first. Couldn't get in? Try submitting to a Tier 2 journal. Couldn't get in? Try submitting to a tier 3 journal. Couldn't get in? Try submitting to a...

Sometimes, you are unsure whether a journal would be interested in your type of results or research. It happens, especially if your field falls across research boundaries (e.g. *bioinformatics* or *biomechanical engineering*). You could then send a pre-submission letter[16] to the journal editor. Most journals even put the format of such a letter on their websites. In that letter, you would present a short brief of your research and contribution, and ask whether it is of interest to the readers of the journal. Two things can happen, both of them good for you. One, the editor politely declines your offer and suggests an alternative journal which may be more appropriate. Two, the editor expresses moderate interest, but mentions areas which would increase his or her interest. Be aware, however, that not all journals accept pre-submission letters.

Now that you've taken all of this into consideration, you should be able to get your paper past the editor's desk and in front of reviewers. It should all be smooth sailing from here, shouldn't it?

The Halo Effect and Confirmation Bias

Unfortunately, we must begin this paragraph with a somewhat depressing but sobering truth. Pull out your manuscript, and gaze upon it lovingly. Have you not spent countless hours writing, rewriting and re-rewriting this document? You know that this finely tuned text into which you've poured your blood, sweat and tears will have judgment passed upon it by the reviewer.

They'll at least spend a solid hour or two combing through it before coming to a decision, correct?

...correct?

As trainers who have taught scientific writing skills for more than two decades, we've had thousands of scientists from all fields come through our workshops. Inevitably, many have had experience reviewing papers. Course after course, we have asked this group of reviewers a simple question: "How long does it take you, on average, to make up your mind about whether or not a paper is worth publishing?" There was of course a range of answers. Some reviewers spent large amounts of time reading a paper before coming to any

---

[16] https://www.aje.com/arc/how-to-write-presubmission-inquiry-academic-journal/

conclusion about its value. However, those reviewers were the exception to the rule. The majority spent an average of only 15 minutes.[17]

If you find yourself thinking that it is impossible to review a paper in such a short amount of time, you would be correct. In fact, 15 minutes is insufficient to read through the whole manuscript. Logically, this would imply that reviewers are deciding whether or not your paper is publishable based on incomplete information. To understand why this occurs, we must foray into the field of psychology.

Reviewers take the responsibility of vetting new contributions to science seriously. But they are also human, and so like any of us, they suffer from evolutionary cognitive biases. The first of these, **the halo effect**, helps explain why they confidently decide whether or not a paper is publishable after just 15 minutes of reading.

Although the term "halo effect" is relatively recent, having only been coined and scientifically proven in 1920[18], the effect itself has been intuitively exploited since the dawn of civilization. Have you ever noticed how in ancient times benevolent deities were portrayed as beautiful, and evil spirits as repulsive or scary? Or in a modern context, have you noticed that the main characters of your favourite TV shows tend to be attractive individuals? Neither one of these scenarios are coincidental: they demonstrate the halo effect at play. We associate positive attributes such as intelligence, creativity and charisma with attractiveness, and negative characteristics with unattractiveness[19]. Objectively, such extrapolations from appearance to character are completely baseless; yet, humans consistently prove that such extrapolations exist and are made subconsciously.

The halo effect does not relate simply to people and their appearance. On a larger scale, it dictates that if our initial impression of someone *or something* is positive, that impression bleeds into other attributes of the targeted person or object[20].

---

[17] Remember, this is an average. Some reviewers responded that it took them only 5 or 10 minutes!

[18] Thorndike, E. L., "A Constant Error in Psychological Ratings," 1920

[19] As opposed to the halo effect's positive associations, negative associations are called the *horn effect*.

[20] A famous example of the halo effect can be seen in the trial of American sports star Orenthal James "OJ" Simpson for the murder of his wife. He was so loved as a

What does this have to do with scientific writing and reviewers? Whereas we make conclusions about someone's appearance instantaneously, getting a first impression of a manuscript takes a little longer (let's say ... *15 minutes on average*. Sounds familiar?) Once the impression has been made, the halo effect dictates that even without having read the rest of the manuscript, the reviewer will attribute to it positive or negative qualities. If the reviewer is impressed by the clarity of the abstract and the significance presented in the conclusion, he or she may subconsciously assume that the methodology is robust and the introduction is sufficiently comprehensive (although again, there is no clear reason to believe that this is objectively true). The reviewer has already decided to "like" the manuscript overall, even without comprehensive reading.

Once that preference has set in, the second cognitive bias comes into play: **confirmation bias**. Having initially felt the paper was worthy of publication, the reviewer will continue reading it focusing on the positive elements of the paper and minimizing its faults. On the other hand, if the reviewer felt the paper was not ready for publication, he or she would then comb through it with the aim of pointing out every little mistake or lack of methodological detail.

Taken together, the halo effect and confirmation bias act as a powerful synergistic mechanism that guides the reviewer's positive or negative response to your paper. Knowing that this process only takes 15 minutes on average, you must make certain that the parts the reviewer reads first, the *critical parts of the paper*, are well written. What are these critical parts?

The title, abstract, structure, introduction (if the reviewer is a non-expert), conclusion, and the visuals (particularly the diagrams in the methodology section and the figures that closely support your contribution). Given how essential these sections are in setting the reviewer's expectations, we have given each one its own chapter in this book.

The halo effect is neither friend nor foe. It is simply a psychological reality that your reader will face. **Will you let it affect you detrimentally, or write in such a way as to put it to good use?**

---

sports idol that many Americans initially supported his innocence, despite these two elements having no relationship to each other.

The Assumption of Expertise Trap

Something in the last paragraph may have puzzled you. We wrote *"if the reviewer is a non-expert"*. Why "if"? Isn't the reviewer always an expert? This may come as a surprise, but it is unlikely that your reviewers are all subject experts. But if they aren't, why have they agreed to review your paper? The selection of a minimum number of reviewers is the job of the editor and associate editors. In order of priority, your reviewers are likely to come from three sources.

- Reviewers who have worked for that journal before, and who are a good fit for the keywords of your paper. Note that the fit will never be perfect! The editor may want to select several reviewers to cover different aspects of your paper. Methodology experts may not necessarily be expert in your specific domain but will evaluate your approach and methodology fairly. Likewise, some "big picture" domain experts will have a better understanding of the significance of your work, but may not necessarily be experts in the approach you chose.
- The list of authors you cite in your references, even if they have not yet been reviewers for that journal. They may be sought out as long as their paper is well cited, and they are seen as being able to comment on the validity of your approach/methodology (but not necessarily on its field of application!)
- As a last resort, the list of reviewers you have suggested in your cover letter or online. Even though these suggested reviewers might be better than some of those chosen by the journal, the editor will only consider them if they do not gather enough reviewers from their usual sources. Since the editor may not know why you propose these reviewers, it would greatly help if you mention why they are particularly qualified to help in the review. Remember that the reviewers you propose cannot have been involved in projects or publications with you in the last three years (or more) and must not belong to the same institution.

Once potential reviewers are identified, the journal will send them your manuscript's title and abstract, and ask whether they are interested in giving their feedback. The reviewers may choose to decline the review if the topic does not interest them or if they are too busy. After all, if they do accept, they will have to spend a considerable amount of time critiquing the paper *for absolutely no pay.* Most scientists agree that the peer review process is critical to maintaining the integrity of scientific practice. But there are also more pragmatic reasons to serve as a reviewer. Being a reviewer gives one the opportunity to read about the latest findings in a field before anyone else (as the publishing process can take 6 months from submission or more to complete). The insights gleaned from the review could yield a competitive advantage in research or save the reviewer's time on preliminary experiments.

Remember how we asked the reviewers in our classes how long it takes them to make a decision about an article? We polled the same group of participants to ascertain if they were always full experts on the topics they agreed to review. Shockingly, only roughly a third of the participants said that they were. This demonstrates that a great deal of non-experts or partial experts[21] are reviewing papers for journals!

Under the highly generous assumption that 80 % of your three reviewers are full experts, what are the odds that at least one of them is a non- or partial expert? We could express this problem mathematically as "what are the odds that not all three reviewers are full experts?" or [1 − (80 % * 80 % * 80 %)]. The answer is 0.488, or 48.8 %. Does knowing that there is a nearly one-in-two chance of having a non-expert reviewer change the way you should write your paper? Can you afford to be as technical as you originally envisioned? Can you afford to assume that the reviewer will understand the assumptions you have made in your experiment or model?

Do not assume your reviewer is a full expert—your identical scientific twin. Write also for the non-expert reviewer.

The Editor

Up until now, we've focused on reviewers as the key readers of your paper. But what about the editor? Arguably, the editor is the most

---

[21] A partial expert is non-expert in some areas of your paper and expert in others.

important reader your paper will ever have. If this one reader decides that the paper is not worth reading, no other readers will ever get the chance to evaluate it for themselves (not even reviewers!). So who is this mysterious person, what does he or she care about, and how can you use the halo effect to your advantage?

The first step to understanding how to please editors is to understand their role. While in your eyes editors may appear to be at the top of the publishing food chain, they too have responsibilities and a body to which they answer. Editors have a mission to accomplish. The better you help them fulfil that mission, the more likely they are to accept your paper. So what is that mission?

Unceremonious as it may sound, scientific journals do not exist solely for the dissemination of research. Journals are businesses built upon the dissemination of knowledge, but they remain first and foremost businesses. They do not receive public money, rarely operate as nonprofits, or act selflessly (why aren't reviewers, an essential cog in the publishing process, paid?)[22]. The editor's job is to make sure that the papers accepted by the journal are of a high quality and likely to garner citations. The more citations, the higher the impact factor for the journal. The higher the impact factor, the more reputable it appears. This reputation translates into a larger source of revenue from a growing readership: individual and academic subscriptions. And finally, the enlarged readership also enables the sale of increasingly valuable ad space. In short, more high quality science = more money.

Fortunately, the publishing industry's goals mostly[23] align with those of researchers: both want to see high quality research published and shared, regardless of the differences in their underlying

---

[22] Although this is generally the situation in 2020, changes in the industry and in regulation are increasingly aiming to remove the money factor from scientific publishing. A handful of journals have started paying reviewers a token sum for their time ($80-$100). Regulation-wise, you may have heard of the Council for the European Union's decision that any EU publicly funded research needs to be freely available from 2020 onwards.

[23] "Mostly", because journals are not as interested in publishing negative findings, leading to significant wastage of research time and money into avenues of research which may already have been found to be dead ends (but were never published).

motivations. But this shared vision only extends so far. For the researcher, scientific merit is the top priority. For the editor, readership takes the spot. How does caring for the readership differ from caring about scientific merit?

The first consideration of the editor is the length of each journal issue. Depending on the amount of money available from advertising, each printed journal issue is limited to a set number of pages. If you submit an article that is considerably long, the editor must decide whether it makes more sense to publish your paper or two other shorter papers, each of which could satisfy the needs of the journal's readers. **An increase in article length equals a decrease in editor appetite. So keep your papers as short as possible without compromising on clarity**. Look through the last issues of the journal and take note of the average article length. Aim to write at that length, or slightly shorter. Alternatively, you may choose to adopt a format that corresponds to a shorter publication, such as a letter.

Secondly, the editor must consider the paper's target audience. Let's imagine you've submitted a paper to the Journal of Nephrology[24]. Who reads that journal? If you thought of nephrologists, you are correct. But nephrologists are only a subset of the total readership. Who else reads the journal? Think about kidney disease researchers, dialysis machine manufacturers, membrane research scientists, patients with a specific kidney disease (and their families), drug manufacturers, etc. The editor's job is to be mindful of these various readers and publish a range of articles that cater to their diverse needs in each issue. **If you can present your work as valuable to more than just a subset of the journal's readership, you multiply the worth of your paper in the editor's eyes.**

Fair enough, you might think, but where can you plead your case to the editor? No journal article starts off with a paragraph in the introduction stating *"We believe the research conducted in this paper would be of value to membrane specialists and sufferers of Nephritis..."*

---

[24] The medical specialty dealing with kidney function and diseases

The cover letter

Writing the cover letter is often the very last step in the months-long process of writing a manuscript. Some even see it as a necessary, but peripheral and unimportant task. What a mistake they are making! Whereas reviewers build their first impression from browsing the manuscript, the halo effect starts to affect editors before any browsing begins. Their first impression, and a strong one, is built while reading the cover letter.

Many cover letters are written in the style of a scientific paper. In fact, we've seen many letters that are word-for-word copied and pasted from different parts of the paper! One sentence from the introduction here, and a few from the abstract and/or conclusion. How do you think the editor feels when they stumble upon some of these same sentences just minutes later while reading the abstract? The image the author portrays by writing the cover letter from cannibalized parts of the paper shows a lack of care towards the reader. And when editors are frustrated by the cover letter, the halo effect whispers that this expediency may also permeate other parts of the paper (even if it does not). **Remember, the editors are the most important readers of your paper. You cannot afford to frustrate them.** So aside from the earlier copy-paste behaviour, what are other common pitfalls that frustrate editors?

1. *The cover letter is incomplete.* Every cover letter will follow the same general structure, but blind reliance on a template can lead to trouble. In their instructions to authors, each journal details what elements the cover letter should contain, including some mandatory statements (e.g. *this paper is not under consideration by any other journal,* etc.). Make sure none of these parts are missing or formatted differently from the instructions.

2. *The cover letter isn't formatted as a letter.* Without reading a single word, you can easily tell a letter apart from a report due to its distinctive format. The first words should be "Dear editor" or "Dear Dr. ____". You should also end the letter with a closing line such as "Thank you for considering our manuscript, we look forward to your response", and a signature

line such as "Kind regards". Finally, you should sign off with your name, title, and institution. If sending the letter offline, also include the traditional elements found on a letter, such as the addressee name and mailing address.

3. *The cover letter isn't written in the style of a letter.* Imagine you're writing a letter to your spouse's grandmother to thank her for buying you a ticket to a broadway musical. Would you start off with "The most thankful of wishes must be imparted upon you, dear reader, for the gift of tickets which were joyfully received by the authors of this communiqué"? Or would you be more casual, in the style of "Dear Lizzie, thank you so much for buying me that ticket, I was extremely happy to receive it!". The cover letter is a *letter*. It is a personal correspondence between two people, not an impersonal document. It should therefore be written in a more conversational style (but not too informally either! Spouse's grandmother, not college friend!). This means that personal pronouns must be present: "we", "our", or if you are the sole author, yes even "I" or "my"! It also means that sentences should be written in the active voice — *"I am writing to submit my manuscript"*, not *"this letter is for the submission of the attached manuscript"*.

4. *The cover letter is too long.* Unless absolutely necessary, the cover letter should fit within one page. You must be selective in your wording and succinctly encapsulate the context for the work, what is new and different, and the significance of your work for the field.

5. *The cover letter is tiring to read.* When was the last time you used acronyms in a personal letter? Neither they nor jargon should belong in the cover letter unless well-defined and absolutely necessary. Only provide necessary technical details if they are essential to telling the story of the research. Remember that the purpose of the cover letter is **not** to convince the editor of the quality of your science or the breadth of your knowledge — it is simply to help them see the manuscript's value and decide whether or not it is appropriate for acceptance and review.

Learn from Principles, Not Examples

I have claimed that learning from examples is poor practice. To justify this claim, I argued that given the poor level of scientific writing, an average paper would make a poor exemplar of how to write well. But let's now imagine that you are fortunate, and that you have a mentor who has earmarked a list of well-written papers for you to study. Do you now believe you have enough to succeed?

Let's briefly consider this situation in terms of a cooking analogy. If, while cooking, you follow a recipe book to the letter, using all the proper ingredients, temperatures, cooking times, and equipment, can you achieve a great-tasting dish? Yes. But does following a recipe give you an understanding of how flavour profiles interact, how sweet and sour tastes combine masterfully, or when to use a specific herb or spice to enhance a type of dish? No. There is an immense difference between an excellent cook and a chef. Between someone who can apply the rules and someone who understands them intimately.

Writing is no different. While one can learn to write well-structured sentences from good examples, more effort is required to deeply understand the rules of writing and apply them strategically. You might observe, for example, that the passive voice is used far more rarely in well-written papers. Subsequently, applying that finding to one's own papers should, in theory, increase their similarity to the well-written exemplars. But an overall reduction of the passive voice, without an understanding of when and how it remains useful, could backfire in key areas. Copying from good papers may address the symptoms of bad writing, but it does not bring you closer to understanding the underlying problems, or to finding a cure.

In order to become a chef (in writing), you need to understand the fundamental principles of writing and reading, and build up your knowledge from there. Doing so clears away any false assumptions you may have gained during your education or work (sadly, many plague the scientific writing style, as we will see later in this book).

For writers, what are these fundamental principles? As the job of all writers is to be understood by the reader, the principles are reader-centered. They include understanding the physiological limitations of any reader, such as limited memory, attention, motivation, or patience; the limitations of the reader's existing knowledge, or the limitations linked to the serial process of reading itself and its dependence on expectations. All of these will be covered in this book.

Flowery, grandiose, and abstract writing may temporarily sound impressive, but it often fails to accomplish the most basic task of communication: to help the reader understand. Great writing eschews the need to impress the reader and focuses on getting its message across clearly.

Chapter 3

The Scientific Writing Style

Do you have a favourite author? Personally[1], I am a big fan of J.R.R Tolkien. As a child, I would stay up at night reading *The Lord of the Rings*. Tolkien was a master at describing imaginary worlds. He could describe a single fictional location over two full pages, breathing life into every little corner, creature, shadow, or stone present. He was so detailed that I felt as though I wasn't only imagining the places in his books, but could almost see them for myself. My brother, on the other hand, wasn't as much a fan of Tolkien's writing. He enjoyed the story and the characters, but the nearly-excessive level of detail got in the way of his reading enjoyment.

Every author has their own writing style. Tolkien does not write in the same way as John Grisham, Ursula K. le Guin, Seth Godin, George R.R. Martin or J.K. Rowling. They don't just tell different stories. They tell stories differently. If this group of authors were to accept a writing challenge where every element of the plot had already been laid out in advance, we would nevertheless end up with six different books. **Do not confuse content with style: content is *what* is described, whereas style is *how* it is described.**

In fact, you do not need to be an author to have a writing style. Individually, each one of us thinks differently, and we thus express ourselves differently on paper. Some, like in this example sentence, place many commas in their writing, leading to frequent pauses in reading. Others prefer punctuation like semicolons or em-dashes. For some people, writing style is affected by a lack of vocabulary — perhaps because English is a second language. Writing style is so individual that a science has developed around its classification and identification: stylometry. Much like each fingerprint is unique, a writing style

---

[1] Co-author Justin Lebrun here!

can serve as a signature identifier. At first, the science was limited to experts in linguistics who had deeply studied specific authors. But with increasing developments in computer-aided techniques and machine learning, its ease of use, precision, and field of applications has rapidly expanded[2].

For the first part of your life, finding your own 'voice', your own writing style, was encouraged. Your teachers in primary and secondary school would make sure that your writing was grammatically correct, but they would also say "Don't just copy what you read. Find your own voice!". You took that advice, and applied it when writing your University or College admission application. Congratulations! You were accepted. You were then shepherded through to the next step of your educational journey, and just as you started getting excited about how your writing would further progress, it stopped. What happened? You encountered a great barrier to individual expression: the academic writing style. The celebration of uniqueness that existed in high school is gone. You had entered into a 4-year cycle that slowly but gradually stripped away individual expression in favor of crafting a standardized product, the academic paper. Personal style was not encouraged; referencing appropriately was. Following the same guidelines as everyone else was. The underlying logic behind this narrowing of styles seemed sound: *a standardized paper will be easier to read because fewer variations from the norm will create fewer reader problems.* Fair enough. But what happens when the norm itself is plagued with problems?

Characteristics of the Scientific Writing Style

Let's take a look at a simple english sentence, written in a casual style:

I wondered if I should go shopping.

What a short and pragmatic sentence! It clearly and concisely expresses the subject and the action of the sentence. But it lacks...

---

[2] Stylometric tools can even be used to identify the authors of a piece of computer code. Each individual's unique writing style clearly applies beyond simple prose!

flair. How would Shakespeare, the great literary author, express this sentence using his trademark style?

To shop or not to shop, that is the question.

Same content, different style. It's interesting, but we're unlikely to use any of Shakespeare's now-outdated phrasing in our own writing. So how would a scientist express it?

`Consideration was given to the possibility of departure for the acquisition of merchandise.`

Did you chuckle to yourself? If so, you should perhaps pause to reflect on why you did. This sentence is so over the top that it is practically parodic. But at the same time, it is difficult to deny that it indeed sounds "scientific". As the adage goes, *it's funny because it's true.*

This sentence expresses the exact same content as the first, simple one: *I wondered if I should go shopping*. But it does so in such an indirect style that you can almost imagine the lab coat. What makes it so "scientific"? Or in other words, what are the characteristics of the scientific writing style? Here are six telltale signs.

Longer sentences: The new sentence is considerably longer than the original. In the casual style, the sentence is 7 words long (35 characters). In the scientific style, the sentence is 13 words long (91 characters). That's nearly twice as many words and nearly three times the length of the original! Research around readability has demonstrated that sentence length is inversely correlated with clarity[3]. Therefore while the scientific style might seem more elevated than the casual style, it is markedly less clear.

Intensive use of the passive voice: "Consideration was given". Who gave the consideration? Is it the author? A committee? As the writing style removed the subject from the sentence, it is impossible to give an answer.

By putting the sentence in the passive voice, the author de-emphasizes the subject's role in the action. *"It does not matter*

---

[3] https://en.wikipedia.org/wiki/Readability#Popular_readability_formulas

that it was me who considered it — what matters is that it was considered". Such de-emphasizing can be appropriate in a scientific paper, especially in the methodology section where it doesn't matter who titrated the solution as long as the task was done. But the unnecessary use of the passive voice, as in this example sentence, is pointless and makes for less dynamic reading. Instead of "who is doing what" (dynamic), all we are left with is "something is being done", or "something results from something else" (passive).

Supremacy of nouns over verbs: Let's count the <u>nouns</u> and *verbs* in this sentence:

<u>Consideration</u> *was given* to the <u>possibility</u> of <u>departure</u> for the <u>acquisition</u> of <u>merchandise</u>.

The sentence contains five nouns, but only one verb (that is weakened by being in the passive voice). Nouns serve to identify people, places, or things — in other words, objects. Verbs bring action to a sentence. With this imbalance of nouns over verbs, many things are being described but with very little action. The sentence lacks life and interest.

Multisyllabic nouns: Not only is the sentence long, but it contains many complex words. Every underlined noun in the sentence above has at least 3 syllables. Why write *merchandise* which has four syllables when you can write *goods?* Or *acquisition* instead of *buy?* The English language is remarkably rich in redundancies, offering its writers many ways of expressing the same idea. Have you ever wondered why this is so?

To answer that question, we must jump back through time nearly a thousand years, and take a closer look at events transpiring in medieval England. In the year 1066, Norman king William the Conqueror invaded the British Isles, establishing an Anglo-Norman[4] speaking government in England. In his generosity (and wisdom), William did not attempt to ban the use of the English language in its current

---

[4] Anglo-Norman is a close relative of French. The written forms of both Anglo-Norman and French are considered *mutually intelligible*, meaning that if you have knowledge of one language, you could fully understand the other without any further education or training.

iteration (Anglo-saxon, or *Old English*). However, he did insist that French would be spoken in his court. Therefore, if you were of the aristocracy, you learned French. If you were educated or wealthy, you learned French. *If you studied in the sciences, you learned French.* Over hundreds of years, Anglo-Norman and Anglo-Saxon combined to form new languages that used words from both vocabularies: Middle English, and eventually Modern English. The integration of French into English was so pervasive that today, it is estimated that 45% of all English words have a French origin.

Let's take another look at the long words from the example sentence. How many of them are originally French? All of them. *Consideration* comes from the French word *considération*. *Possibility* from French word *possibilité*, *departure* from French word *départ*, *acquisition* from the identically-spelled French word, and *merchandise* from French word *marchandise*.

We find it personally fascinating that the scientific writing style is in fact a French saturated style. Objectively, these French-based multisyllabic nouns decrease readability. So why do scientists still today, in 2020, use such vocabulary? As we have seen, academics of the eleventh century had to learn French to ingratiate themselves within the upper ranks of society. This linguistic tradition was passed down century by century, surviving even today. Could it be that the reason modern scientists write unclearly is due to historical events that took place nearly a millennium ago? Does it seem a cultural legacy worth preserving, if it directly comes into conflict with reader understanding? As always, we encourage you to make a reader-centered decision. It is more important to be clearly understood than to write in a style that "seems scientific".

Specialized vocabulary: Generally speaking, multisyllabic nouns are more rarely used than their short saxon counterparts, but not always. The French-based word *continue*[5], for example, is widespread in English. The scientific writing style not only suffers from having many multisyllabic nouns, but tends to use specialized ones that are otherwise rarely used in daily conversation. Unless working in a logistics or freight company, the word merchandise is rarely

---

[5] The saxon equivalent of *continue* is to *go on*.

used. And when was the last time you found yourself speaking the word *acquisition* out loud? On top of that, we're just examining an example about shopping. If we plunge into the actual jargon of scientific fields, it is almost comparable to having to learn a whole new language. One of the reasons that scientific papers are especially difficult for younger readers is that they have to grapple with building both knowledge *and* vocabulary simultaneously, a gruelling task.

Hedge words: As this is a relatively obscure grammatical term used in a subfield of linguistics, let us clarify what hedge words are using simple examples. In sentences without hedges, things tend to be black or white, with little room for interpretation: *he is the best boxer in the world.* Introducing a hedge word removes that certainty and replaces it with possibility: *he **might** be the best boxer in the world.*

Hedges can be adjectives (*possible*), adverbs (*potentially*), verbs (*seem*) or even phrases (*it is likely that*). In scientific writing, it is essential to hedge your claims as they might otherwise be disputed or overstated. At the outcome of a study showing a correlation between air pollution levels and hair loss, it would be dangerous to claim that *"air pollution is responsible for hair loss"*. A reviewer could point out that other factors or variables which have not been controlled for might instead be responsible. If, however, you write *"air pollution **seems** to increase hair loss"*, a reviewer is less likely to argue.

Concluding on the previous paragraphs, we have seen that the scientific writing style tends to favor long passive sentences composed mostly of specialized multisyllabic nouns and hedge words. Independently, none of these characteristics are inherently bad. There's nothing inherently wrong with using hedge words. But when these words are overabundantly applied or used incorrectly, writing suffers. There's nothing wrong with a sentence in the passive voice. But when multiple passive voice sentences stack up, the reader loses all interest. Combine the excessive use of the passive voice with excessive hedges, and writing suffers even more.

We are not asking you to abandon the scientific writing style altogether. But we are making you aware of its weaknesses, so that you may decide whether it is helpful or harmful. Do not be shy to deviate from the norm! While you may have been graded on your conformity to a standard of academic writing in your tertiary education days, the real evaluators of your scientific paper are its readers — your peers. We assure you that the majority prefer clarity and ease of reading over convention. Wouldn't you feel the same way?

Understanding Sentence Length

Long sentences may cause readers difficulties. But it doesn't quite feel fair to lump all long sentences together. Some very long sentences remain easy to read while other slightly shorter sentences are impossible to get through.

Clauses

Let's take the next two sentences as an example:

General Sanders had never met a soldier like the new recruit from Delaware, Private Jonas Jameson, whose belligerent attitude was evident in every little undertaken action, from guard duty, to report writing, to any and all other administrative tasks. (Long but clear)

Although he knew how to cook well, due to his being stuck in a kitchen whose windows were jammed shut and whose airflow suffered due to a broken ventilator, he found himself distracted and unable to concentrate on the dish. (Long but difficult to read)

What distinguishes one sentence from the next? To understand the answer, we can't just look at sentence length from the perspective of word count or even word length. Both of the sentences above are roughly the same length (39 vs 36 words), and neither one contains many long multisyllabic words. We need another metric by which to evaluate these two sentences: the clause.

According to the Oxford dictionary, a clause is "a unit of grammatical organization next below the sentence in rank and in traditional grammar said to consist of a subject and predicate". Unless you're a big fan of reading books about grammar, that explanation is likely clear as mud. To simplify, a clause is a part of a sentence that contains a subject and a verb.

For example, "*He met his wife*" is a sentence made up of a single clause — only one subject, *he*, and one verb, *met*. We can extend that sentence by turning it into a two-clause sentence: *He met his wife while she went shopping*". We now have two pairs of subjects and verbs — *he met* and *she went*.

Notice that although the first clause could stand on its own (*He met his wife* is a grammatically correct sentence), the second clause cannot stand alone (*when she went shopping* cannot be a full sentence). We call the first type of clause that stands alone a **main clause** and the second type a **subordinate clause** (or subclause).

He met his wife when she went shopping

You could further elongate the sentence by adding a third subclause: *He met his wife while she went shopping because she walked into his workplace.* Or even a fourth: *He met his wife while she went shopping because she walked into his workplace which was located next to the town's public library.* We've come this far, why not add a fifth? *He met his wife while she went shopping because she walked into his workplace which was located next to the town's public library that was under renovation.* Notice how even though the content of the sentence is simple, it is beginning to become more and more difficult to read. Why? Because it contains many different subjects and many different actions, each one of which gives the reader some information:

1) He met his wife
2) This meeting happened while she was shopping
3) You can shop at his workplace
4) His workplace is located next to the town's public library
5) The library was under renovation

Trying to piece together these disparate pieces of information within a single sentence is tiring for the reader, and may require a second read. Even though the sentence isn't particularly long (27 words), it already *feels* long because of information density.

Do a little exercise and try to identify the number of clauses present in the two long sentences proposed right under the earlier heading "clauses". Look for subjects and verb pairs. When you're done, continue on below for the solution:

—

General Sanders had never met a soldier like the new recruit from Delaware, Private Jonas Jameson, whose belligerent attitude was evident in every little undertaken action, from guard duty, to report writing, to any and all other administrative tasks.

Despite its length, the sentence only contains two clauses. Now let's look at the second example:

Although he knew how to cook well, because he was stuck in a kitchen whose windows were jammed shut and whose airflow suffered due to a broken ventilator, he found himself distracted and unable to concentrate.

Here, we can see that the sentence's length comes from no less than five clauses!

One clause is easy. Two are fine. Three may require some effort. Four is taxing. Five and above? You're exhausting the reader. If you want to avoid burdening the memory of your reader, **keep to an average of two clauses per sentence**.

Prepositions

Prepositions are amongst the most common words in the English language, appearing in nearly every sentence, often multiple times. You'll recognize them instantly: *for, to, with, about, at, under, through,* etc. How are they used? Their purpose is to add detail to whatever they follow in a sentence. For example, "the man smiled" is less precise than "the man <u>with</u> the balloon smiled", which is itself less precise than "the man <u>with</u> the balloon smiled <u>at</u> the mirror. But adding too many prepositions in a single sentence is like adding detail to detail to detail until the reader is lost. Let's look at a poorly written sentence with too many prepositions:

Researchers have observed an increase <u>in</u> the amount of energy generated <u>by</u> the splitting <u>of</u> uranium atoms <u>in</u> simulations <u>with</u> constraints removed.

Notice how you could place a period before every preposition and the sentence would still make sense. Each preposition simply elongates the sentence with additional detail. If you remove all meaningful information, you can see the bare sentence structure:

"A have observed differences in B of C of D in E with F". Without even reading the sentence, you can already tell it is overly complex. Let's take another look at the sentence, but rewritten to remove excessive prepositions:

Researchers have observed increased energy generation <u>when</u> splitting uranium atoms <u>in</u> unconstrained simulations.

With only two prepositions left, the text is easier to read. Let's take a look at the different ways you can remove excessive prepositions from your text:

Compound nouns: The collector <u>of</u> books → The book collector

Possessive case: The owner <u>of</u> the jeep → The jeep's owner

Passive to active: He was booed offstage <u>by</u> the crowd → The crowd booed him offstage

Use adverbs: He cleaned the floor <u>with</u> great vigor → He cleaned the floor vigorously

Use precise verbs: He looked <u>at</u> the wall carefully → He inspected the wall

The Scientific Style Virus[6]

Take a look at the following sentence:

To get a ballpark figure, the research team, aided by their collaborators at the German Health Policy Institute which they partnered with the year before, adopted the maximized survey-derived daily intake method for the evaluation of the per capita intake of the monosaccharides.

---

[6] Jean-Luc Lebrun, one of this book's authors, has published a free book on the topic. You can download it here: https://www.researchgate.net/publication/335689539_Scientific_Writing_Style_Virus or here https://books.apple.com/us/book/scientific-writing-style-virus/id1476877645

If reading that sentence made you feel a little ill at ease, congratulations, you are a healthy reader. If however you thought to yourself, "my, what a wonderfully written sentence!" I'm afraid I have bad news for you. The sentence above is plagued with errors common to the scientific writing style: word haze, sentence meanders, deactivated verbs, and word spikes! If you felt comfortable in their presence, it is because you have been infected by the scientific style virus. Your writing may be plagued by its symptoms, and your prognosis is poor. But don't despair! The disease has not yet progressed to its terminal stages, and it's nothing a little diagnosis and writing therapy can't fix.

Sentence Meanders

To get a ballpark figure, the research team, **aided by their collaborators at the German Health Policy Institute which they partnered with the year before,** *adopted the [...]*

Let's diagnose the first symptom of the scientific style virus: a meander in a sentence. Here are three versions of the same sentence, one of which is less readable than the others. Which one and why? For now, don't focus too much on analyzing the sentence, just go with your gut feeling.

Because he feared being caught with a fake license that had been purchased in an online marketplace, the drunk driver fled the scene of the accident.

The drunk driver, who feared being caught with a fake license that had been purchased in an online marketplace, fled the scene of the accident.

The drunk driver fled the scene of the accident because he feared being caught with a fake license that had been purchased in an online marketplace.

—

The least readable sentence is the second one. Why? Let's visualize all three using the earlier example of standing and kneeling men to represent main clauses and subclauses.

Through decades of reading, readers have been trained to expect that a subject is soon followed by a verb. When the subject and verb are separated by too great a distance, the reader needs to lock up precious memory resources to keep the subject in mind until a verb

Because he feared being caught with a fake license that had been purchased in an online marketplace, the drunk driver fled the scene of the accident.

Because he feared being caught with a fake license that had been purchased in an online marketplace, the drunk driver fled the scene of the accident.

Because he feared being caught with a fake license that had been purchased in an online marketplace, the drunk driver fled the scene of the accident.

is found. While the reader hunts for the verb, anything else read is deprioritized.

Free up this unnecessary allocation of memory by keeping subject and verb together.

Word Spikes

Ah, Valentine's Day. Love is in the air, chocolates are in ribbon-wrapped boxes, and roses abound in cellophane bouquet sleeves. But these roses aren't like those found in nature — an essential element has been removed from them: the thorns. Rose thorns may not be conducive to Valentine's Day sales, but they certainly are conducive to a rose's survival in the wild. Just as their bright colors and sweet scent attract humans to them, roses also attract animals and insects such as caterpillars which would do it harm[7]. Each thorn serves as a defense mechanism for the rose. To caterpillars, a thorn is fatal if the poor insect impales its soft underbelly on it, stopping its upwards progress towards the petals. Just like rose stems, your writing can also contain thorns — word spikes that impale the minds of readers in their arduous uphill journey to the end of a sentence. Fortunately for writers, the average reader is not as fragile as a caterpillar, and can withstand a spike or two. But as the brain gets stopped again and again by word spikes, it loses much momentum and clarity, and finds itself wondering "what did I just read?" Let's analyze these words spikes one at a time.

Colloquialisms

To get a ballpark figure, the research team [...]

Americans know what a ballpark figure is. But if you're a researcher from another country, the odds of knowing the expression drop dramatically. You know what a figure is, but what is a ballpark figure? Try as you might, you can't make sense of the expression. Is it an image of a ballpark? You head to google image search to look up "ballpark", and the screen fills up with photos of stadiums, raising more questions! Actually, in this context, a ballpark figure means an approximate figure.

---

[7] Arguably, we humans cutting them in great numbers to celebrate our holidays are capable of harming roses more effectively than anything in the animal kingdom. Put another way, humans are the world's most effective rose predators.

In today's world of internationally-shared research, colloquialisms are alienating. You no longer write for readers from your country. Scientific writing needs to be accessible to all[8].

Long compound nouns

To get a ballpark figure, the research team [...] adopted **the maximized survey-derived daily intake method** *for the evaluation of the per capita intake of the monosaccharides.*

We've seen that compound nouns can be useful for removing excessive prepositions, as in turning *the giver of gifts* into *the gift giver.* This removal seems to help clarify the sentence, but could one go too far? *A history of complications in relationships which are romantic* could be shortened to *a romantic relationship complication history*, but such a dense phrase is difficult to digest. It would be better to keep at least one preposition: *A history of romantic relationship complications.* In the example above, *"the maximized survey-derived daily intake method"* could be decompressed into a full, long sentence: *The method which used the maximized values of survey-derived answers about daily intakes* [of monosaccharides]. Compressing the 13 words of the previous sentence into the 5 words of the compound noun is impressive in terms of being concise, but markedly unimpressive from the perspective of clarity. **If ever conciseness and clarity clash, prioritize clarity.** Better for you to be understood at the end than have your reader confused from the start!

To understand why long compound nouns are complicated to understand, I will quote an excerpt from another of our books on writing, *Think Reader*[9], which delves into this topic in great depth:

"Compound nouns often seen together like *dinner plate* or *swimming pool* are unambiguous and easy to understand. Some have even

---

[8] If you're still curious to know the origin of the expression, a ballpark figure means an approximate figure because unlike in football, where every playing field has exactly the same dimensions, a baseball park's dimensions are variable.

[9] *Think Reader* does not focus on the scientific paper, but more broadly on writing itself. All humans have physiological shared limitations when it comes to reading: limited attention, limited time, and limited memory. *Think Reader* explores those limitations, and how good writing can bypass them. Think Reader is available on iBooks, Kindle and Amazon.

become single nouns like *firewall* or *toothpaste*. When compound nouns have more than two nouns, ambiguities arise. "Bomb threat" and "threat detection" combine to form bomb threat detection. But where do you put the invisible brackets? Between [bomb] and [threat detection], or between [bomb threat] and [detection]? To clarify the phrase, add the preposition of: the detection of bomb threats, not the threat detection of bombs. For a reader, left-to-right bracketing that follows natural reading order is the easiest to unfold.

[Bomb]
[Bomb Threat]
[Bomb Threat] [Detection]
[Bomb Threat Detection] [Squad]

However, the unfolding can be more intricate as in the "Turin football club," where the brackets are between [Turin] and [football club]: the football club of Turin, not the club of Turin football. The next compound noun, like complex origami figures, has to be unfolded and refolded differently, which takes more brain processing power.

[Australian]
[Australian football] — the football played in Australia.
[Australian football fan] — The reader is not sure whether to unfold as [fan] of [Australian football] or a [football fan] from [Australia]. I chose the [Australian] [football fan].
[Australian football fan club] — The appearance of "club" creates a doubt.

Do I unfold and refold into the [fan club] of [Australian football]? After all, there might be such a thing as "Australian football" with rules differing from those of regular football. Or do I keep my original unfolding and go with [Australian] [football fan club]? It is clearly ambiguous. A quick look online confirmed that Australian football is indeed different. But what if your reader does not spend the time to search online and continues reading? With insufficient background to accurately bracket the compound noun, the chance of misunderstanding is high (50%). As a writer, you cannot afford to gamble on reader understanding."

Latin or Greek

*[...] for the evaluation of the **per capita** intake of the monosaccharides.*
Like colloquialisms, Latin or Greek words can cause readers unfamiliar with the vocabulary to stumble. Why use *per capita* instead of

the english equivalent, *per person*? *Per capita* isn't more concise, or more precise. It means the same thing. Unlike colloquialisms which demand that the reader share a cultural point of reference with the author, the unnecessary use of Latin terminology demands that the reader shares knowledge of a dead language.

I am not denigrating the use of all Latin. Its use in botany, for example, is encouraged. Because a single plant may have many names in different regions (such as a popular vegetable dish called kangkong, a.k.a water morning glory, water spinach, river spinach, water convolvulus, ong-choy, or swamp cabbage), it can be more precise to identify it by its latin name, *Ipomoea aquatica*. But replacing *per person* with *per capita* serves no clear purpose.

Jargon

[...] *intake of the* **monosaccharides**.

Jargon and the scientific writing style go hand-in-hand. Not only is their relationship natural, it is also constantly evolving. As science deals with discovery, researchers need to give names to things, concepts, or elements that were previously indescribable. How does one describe something never before described? How does one choose how to name the unnamed? Some discoveries take on the name of their discoverers, such as *the Dunning-Kruger effect* or *Moore's law*. Scientists may also choose to turn towards Latin and Ancient Greek for inspiration: the material of human nails and rhinoceros horns is identical: keratin, from the Greek word *keras* (horn).

Unfortunately, neither one of these naming schemes is helpful to the reader. For the first scheme to be effective, the reader would have to recognize the name of the researcher and be familiar with their body of work. In a very small or niche field, this could be sustainable... at least at first. But as the field and number of published authors grows, no one could be expected to keep up. The second scheme, Latin or Greek-based naming, is just as ineffective. Were we all fluent in Latin and Ancient Greek, our knowledge would have allowed us to imply the meaning of new scientific terms. But few of us are fluent in dead languages.

With no logical or intuitive way of understanding jargon, readers need to rely on their knowledge and contextual reasoning to understand a passage with jargon in it. Imagine you encounter the following sentence in a text:

The internet is a great resource for researchers and ailurophiles.

You may be unfamiliar with the word *ailurophile*, so you very rapidly attempt to use logic to deduce a meaning to the word. You may recognize that the word ends with "phile", a fairly common latin-based suffix that indicates attraction to something, such as in *audiophile* or *cinephile*. But unlike *audio-* and *cine-*, what *ailuro-* represents as a prefix draws a blank. Perhaps looking at the rest of the sentence will shed some light? *Researchers and ailurophiles* — default behaviour would suggest that somehow *researchers* and *ailurophiles* are logically connected. Is an ailurophile a subtype of researcher who is interested in the topic represented by the (Greek) prefix *ailuro-?* Finally giving up (the previous thought process only took a couple of seconds), you turn to the dictionary and find out that an *ailurophile* is simply a person who loves cats, something that would have been impossible to guess otherwise.

Every time you write with jargon, you add locks to the sentence. Only readers with the right keys, prior knowledge, can access the sentence's full contents. The more jargon you use, the more locks you add, and the higher the likelihood that someone gets locked out.

The jargon of a field should be considered a different language. Casual English is not the same as business English, legal English, medical English or engineering English. Each field brings with it an entire vocabulary of jargon. And much like learning a language, gaining fluency in Scientific English is a task that takes a long time to master. **Should your readers learn a new language — or should you write in theirs? The question has no easy answer**.

We are not advocating to ban the use of jargon — simply recommending that you tailor it to the level of the reader. How do you determine that level? Which publication medium are you targeting? Is it an internal report read only by people in your research team who are intimately familiar with the jargon? Is it a niche journal read only by field experts also familiar with your jargon? Or are you targeting a broader journal where jargon will lock out potential readers and prevent them from using your work or citing you? Finally, if you are aiming for the top tier journals with the widest readership such as *Science*, realize that since their readers come from diverse fields, the use of jargon needs to be minimized, or at the very least, explained as it is introduced.

Acronyms as Jargon

Acronyms are an interesting subcategory of jargon.

The man's BP rose dramatically when his ex-wife entered the room.

Did the previous sentence make sense to you? It would, if you knew that BP stands for *blood pressure*. While any reader would have understood the non-abridged version of the sentence, it is likely that only those in the medical field (for whom BP is a common abbreviation) had full comprehension. Inversely, *I work for International Business Machines* probably doesn't say much to you, but you would instantly understand if I instead told you *I work for IBM*.

Acronyms are useful shortcuts to condense many words into a compact form. For example, using *CRISPR* instead of *clustered regularly interspaced short palindromic repeats* makes sense. But using *BP* instead of *blood pressure to* save one word and introduce a potential lock makes little sense. As the number of acronyms in a text grows, the difficulty in reading grows not additively but multiplicatively. **Acronyms create conciseness, but at a high memory and knowledge cost. Use them only if absolutely necessary**.

Deactivated Verbs

*[...] intake method **for the evaluation of** the per capita intake of the monosaccharides.*

Have you ever felt a tinge of boredom while reading scientific texts? Sentence meanders and word spikes may both decrease clarity, but they shouldn't affect reader excitement. So if not them, what does? As humans, we are attracted and interested by change and action. *He fires critiques* is more exciting to read than *he is critical* because one excites and raises curiosity while the other is a flat description. *He chased and arrested the thief* is more exciting than *arrest of the thief happened after a chase was initiated* because the former focuses on *someone doing something,* whereas the latter focuses on *something that was done*. Let's explore how to raise or lower the reader's excitement in greater detail by analyzing some examples:

Compound crystallization occurs after the reaction between water and sodium has taken place.

While grammatically correct, this sentence is a bit lengthy and lacks dynamism. Here it is, rewritten:

The compound crystallizes after water reacts with sodium.

Notice that the rewritten sentence is considerably shorter (57 characters vs 92 characters) and more action-driven. What differentiates these two sentences? Their verbs.

Verbs can be broadly classified into two categories in English: *action verbs* and *stative verbs*. *Crystallizes*, like *measures, jumps*, or *calculates*, describes an action taking place. *Occurs*, on the other hand, like *remains, takes place*, or even the common verb *is*, does not describe an action but a state of being. When we deactivate the verb *crystallizes* by turning it into the verb-noun pair *crystallization occurs*, we are replacing a strong action verb with a weak stative one. This process of turning verbs or adjectives into nouns has a name: nominalization. To remove nominalizations, adopt the following protocol:

Measurements of the hull were taken by the research team

1) Identify the verbs in the sentence. Are there any weak stative verbs?
2) What nouns do they qualify?
3) Can that noun be turned into a verb? (e.g. *measurements* into the verb *to measure*)

Not all nominalizations need to be removed to breathe excitement and life into a text. It is not their presence but their abundance that makes scientific writing so challenging to read.

Try to rewrite the following paragraph with fewer nominalizations using the technique described above:

Owing to the lack of information on interactions between VR headsets and users over the long term, scientific uncertainty on the occurrence of adverse effects on users due to prolonged usage remains.

—

Because we lack information on how VR headsets and users interact over the long term, we cannot ascertain whether users are adversely affected by prolonged use.

The lack (noun) becomes *we lack* (verb), *interactions* (noun) becomes *interact* (verb), *uncertainty* (noun) becomes *to ascertain (verb), and adverse effects* (adjective and noun) becomes *adversely affect* (adverb and verb).

—

Curing the Scientific Style Virus

We've seen that the virus reveals its presence in meandering sentences, word spikes, and nominalizations. Now it's your turn to move on from diagnosis onto the cure. Try to rewrite the following sentence removing as many viral symptoms as possible. (One potential correction will follow below)

To get a ballpark figure, the research team, aided by their collaborators at the German Health Policy Institute which they partnered with the year before, adopted the maximized survey-derived daily intake method for the evaluation of the per capita intake of the monosaccharides.

1) Remove the colloquialism word spike by replacing *ballpark figure* with *approximate figure*.
2) Notice that *to get an approximate figure* is a nominalization. Replace with *approximate* the verb.
3) Remove the sentence meander about the German Health Policy Institute by shifting it either to the beginning or end of the sentence, or place it as its own separate sentence.
4) Dismantle the long compound noun by introducing a few prepositions.
5) Replace the Latin with simple English.
6) Define the jargon so it can easily be understood.

Through a survey conducted in partnership with the German Health Policy Institute, we approximated the maximum daily intake of simple sugars (manosaccharides) in people's diets.

Chapter 4

Require Less from Memory

Before we jump in, we would like to introduce you to Vladimir - a Russian scientist who will occupy many little story bubbles that you will find across this book. Observing his behaviour is an often humorous way to pinpoint the pains that we all encounter as scientific readers.

Reading accident

Vladimir is reading a paper in the printed proceedings of the conference he attends each year. As it turns out, he is reading your paper! Suddenly, he stops on page three, places his index finger underneath a word, and rapidly scans the text he has just read, searching for something. What he is looking for is not on the page. With his free left hand, he flips back one page, and then another... He stops again. His face lights up. Satisfied, he returns to the page he was reading before the distracting reading U-turn, withdraws the finger still pointing to the problematic sentence, and resumes reading. What happened? A reading accident: the forgotten acronym. Who is responsible? Guess who!

The Forgotten or Undefined Acronym

Acronyms allow writing to be concise, but conciseness is unhelpful if it decreases clarity. An acronym is always clear within the paragraph in which it is defined. If it is used regularly in the paragraphs that follow, the reader is able to keep its meaning in mind, but if it appears irregularly or if reading is frequently interrupted, the acronym — away from the warm nest of the reader's short-term memory — loses its meaning.

How could Vladimir have lost the meaning of the acronym? He had read its definition at the beginning of the paper, but had forgotten it by the time he got to page three. Apart from simply forgetting, Vladimir could have missed the definition all together. Picture this.

You drew a brilliant figure that attracts the attention of Vladimir as he first browses through your paper. He is drawn to the caption of that figure and trips over an obscure acronym. Problem is, Vladimir has not yet read the introduction of the paper where the acronym is defined. Therefore, refrain from placing acronyms in captions and subheadings.

Sometimes, the acronym is not defined at all because the writer assumes that the reader is an expert in the field, and already familiar with the acronym. This omission is sure to upset the readers who are not experts because it forces them to search for the definition of the acronym outside of the paper.

Avoiding problems with acronyms is easy:

- If an acronym is used two or three times only in the entire paper, it is better not to use one at all (unless it is as well-known as IBM, or the paper happens to be extremely short, like a letter or an extended abstract).
- If an acronym is used more than two or three times, expand its letters the first time it appears on a page or under a new heading so that the reader never has to go far to find its definition.
- Avoid acronyms in visuals or redefine them when in a title, legend or caption because readers often look at the visuals of a paper before reading it.
- Avoid acronyms in headings and subheadings because readers often read the structure of a paper before reading the rest.
- Be conservative. Define all acronyms, except those commonly understood by the readers of the journal where your paper is published.
- Never leave an acronym undefined unless it is as familiar as U.S.A. And remember that your international reader may not be familiar with local, national, or foreign acronyms (French SIDA is English AIDS).

> ### The Singapore taxi driver
>
> *The other day, while Vladimir was in Singapore, He hailed a taxi to go to a research institute located on the campus of Nanyang Technological University (N.T.U). The taxi stopped. Vlad got in and said "Nanyang Technological University, please". The taxi driver, an old man who had clearly been living many years in Singapore, replied: "I do not know where that is". The answer surprised Vlad. The university is old and well established; surely the taxi driver had taken passengers there before. When Vladimir explained that it was at the end of the expressway towards Jurong, the old man's face suddenly brightened and he said with a large smile, "Ah! N.T.U! Why didn't you say so before!" That day, Vladimir learned that an acronym is sometimes better known than its definition.*

☀ If the acronym is well-known, introduce it before its definition. If it is not well-known, let its definition precede the abbreviated form.

The popular universal learning algorithm SVM (Support Vector Machine) had a profound impact on the world of classification.
The new universal learning algorithm — Support Vector Machine (SVM) — is likely to have a profound impact on the world of classification.

Identify the first appearance of each acronym in your paper. Search for its subsequent appearances. If the acronym reappears in a section of your paper different from the one where it was first defined, redefine it there. That way, the reader never has to go back further than the head of a section to find the meaning of any acronym. If the acronym appears in a heading/subheading, or in the caption of a figure, replace the acronym with its full definition.

The Detached Pronoun

This, it, them, their, and *they* are all pronouns. A pronoun replaces a noun, a phrase, a sentence, or even a full paragraph. Like the acronym, it acts as a shortcut to avoid repetition.

Pronouns and acronyms are both pointers. This characteristic is at the root of the problems they create:

- If you point to where someone was sitting an hour ago to refer to that person, not even the people who were there may remember who that person was. If the pronoun points to a noun already gone from the reader' short-term memory because it is located 80 words back in the text, the noun-pronoun link is broken. This disconnection is insufficient to discourage readers. They read on, but with less clarity of thought.

- If during your speech you mention and point towards a friend standing in a group thirty meters away from you, only the people who know that person will know whom you are referring to; no one else will. Prior knowledge disambiguates. When the pronoun points to several likely candidates, the non-expert reader — whose incomplete understanding of the text does not allow disambiguation — will pick the most likely candidate and read on, hoping to be right. If that likely candidate is the wrong one, then interpretation errors follow.

   The fat cat ate a mouse. It had been without food for a week.

   What had been without food for a week, the cat or the mouse? The writer knows, of course, but the reader, left guessing, attempts to guess what '*it*' means, using logical inference. One could assume that '*it*' refers to the cat — if it was hungry from not eating for a week, it would make sense that it chased and caught the mouse.
   If instead the adjective "thin" were to be added to describe the mouse, as in "the cat ate a thin mouse", we would then use our logic to infer that *it* in the next sentence now describes the mouse.
- Finally some fingers seem to point to nobody, or they point to things still to come.

   It was deemed appropriate to exclude women from the sample.

   This sentence starts with the infamous '*it*,' the passive voice companion. Who is '*it*'? The person who deemed '*it*' appropriate is hidden from the reader. Is '*it*' the writer? The supervisor? The research department manager?
   And some fingers point to things or people still to come. If you point to where someone is going to be standing five minutes from now to refer to that person, the people will have to wait to know who that person is. This creates tension.

   *Our surprise speaker for this evening is a great archer. **He** has been ....*
   ... Ladies and gentlemen, let's welcome Professor William Tell.

Apart from these situations where the substitute for the pronoun lives in the reader's future, it only takes milliseconds for the brain to choose a likely candidate for the pronoun. What influences the choice of candidate?

1) **The reader's knowledge.** The more superficial the knowledge, the more error-prone the choice will be.
2) **The context.** When there are multiple candidates, the reader uses context and logic — as in the cat example.
3) **The distance between the candidate and the pronoun.** Candidates that are many words away from the pronoun are unlikely to be chosen.
4) **The grammar.** A singular pronoun (this) or a plural pronoun (these, them, they) helps guide the choice, and so does gender (he or she, her, his).

In the following example, try to determine what the underlined pronoun 'their' refers to. The three candidates are in bold. Had the sentence been clear, this task would have been instantaneous. As it is, you will probably struggle; but if you do not, ask yourself how much your knowledge of the field has helped you make the correct choice.

> The cellular automaton (CA) cell, a natural candidate to model the electrical activity of a cell, is an ideal component to use in the simulation of **intercellular communications**, such as those occurring between cardiac cells, and to model **abnormal asynchronous propagations**, such as **ectopic beats**, initiated and propagated cell-to-cell, regardless of the complexity of _THEIR_ patterns.

It is difficult to determine the plural noun pointed to by 'their' because the sentence segment 'regardless of the complexity of their patterns' could be moved around in the sentence, and still make sense.

> ... to use in the simulation of intercellular communications, regardless of the complexity of their patterns...
> ...to model abnormal asynchronous propagations, regardless of the complexity of their patterns...
> ...such as ectopic beats, regardless of the complexity of their patterns...

Communications, propagations, and beats can all display complex patterns. Let us decide that in this text, 'their' represents the 'abnormal asynchronous propagations.'

The ambiguity can be removed in different ways. One could omit the detail. The long sentence would be seven words shorter.

The cellular automaton (CA) cell, a natural candidate to model the electrical activity of a cell, is an ideal component to use in the simulation of **intercellular communications**, *such as those occurring between cardiac cells, and to model* **abnormal asynchronous propagations**, *such as* **ectopic beats**, *initiated and propagated cell-to-cell.*

One could rewrite the part of the sentence that creates a problem to make the pronoun disappear.

The cellular automaton (CA) cell — a natural candidate to model the electrical activity of a cell — is an ideal component to use in the simulation of intercellular communications, such as those occurring between cardiac cells, and to model the cell-to-cell initiation and propagation of abnormal asynchronous events (such as ectopic beats) with or without complex patterns.

One could repeat the noun instead of using a pronoun.

The cellular automaton (CA) cell, a natural candidate to model the electrical activity of a cell, is an ideal component to use in the simulation of intercellular communications, such as those occurring between cardiac cells, and to model abnormal asynchronous events, such as ectopic beats, initiated and propagated cell-to-cell, however complex the propagation patterns may be.

However, these are all quick fixes. The best way to handle this situation is to return to the source of the problem, in this case, the extreme length of this sentence. A complete rewriting is necessary.

Cardiac cells communicate by initiating and propagating electric signals cell to cell. The signal propagation patterns are sometimes complex as in the case of abnormal asynchronous ectopic beats. To model such complexity, we used a Cellular Automaton cell (CA cell).

The length of the rewritten text is only about 70% that of the original. The paragraph is clearer, shorter, and without the pronouns 'those' and 'their'.

Conduct a systematic search for each of the following pronouns in your paper: 'this,' 'it,' 'they,' 'their,' and 'them'.

If you were the non-expert reader, could you easily identify what the pronoun refers to without ambiguity? If you could not, remove the pronoun and repeat the noun(s)/phrase it replaces. An alternate route consists in rewriting the whole sentence in a way that removes the need for the pronoun.

> ### Pattern matching
>
> *That day, Vladimir could not understand why the paragraph he was reading was so obscure. The usual culprits were absent: the grammar was correct and the sentence length was average for a scientific article. Then Vladimir roared as if he had aligned three gold bars in a slot machine. His superior pattern matching brain had aligned three synonymous expressions:*
>
> *(1) Predefined   location   information*
> *(2) Pre-programmed location   information*
> *(3) Known   position   information*
>
> *Each column contained either identical words, or synonyms. Vladimir decided to replace each of the expressions with 'known location'. When the fog created by the synonyms cleared, the structural problems of the paragraph appeared in full light.*

The Diverting Synonym

We all remember (not always fondly) our English teacher from primary school; let's say her name was Mrs Smith. She would read our writing exercises out loud in class, and scold us each time we dared repeat the same noun in successive sentences.

"The fat cat ate a mouse. The cat had not eaten a mouse for weeks."

She would pounce on us and mortify us for having committed the dreaded eighth capital sin: repetition.

"But little Johnny, you should not repeat 'cat' or 'mouse'. You should use synonyms. For example, you could say 'The feline creature had not eaten a rodent for weeks'". And poor little Johnny, who did not even know what felines or rodents were, marveled at the deep knowledge of his English teacher. The message he carried into his life as a researcher was the very message that Mrs. Smith had drilled into him: do not repeat words but use synonyms to demonstrate your great culture through your vast vocabulary.

In scientific papers, however, synonyms confuse readers, particularly those not familiar with the specialized terms used in your field.

☀ Avoid synonyms. Make your writing clear by consistently using the same keywords, even if it means repeating them from óne sentence to the next. As an added benefit, you will lessen the demand on the memory of your readers: fewer synonyms are fewer words to remember and understand.

Search for synonyms. They are usually found in successive sentences. Synonyms of title keywords are often found in abstracts, in headings and subheadings, and in figure captions. Select your set of keywords (particularly those used in the title) and reuse these keywords everywhere: title, abstract, introduction, headings and subheadings, legends, and captions.

The Distant Background

> ### The Macintosh factory
>
> *When Vladimir moved to Cupertino California, in 1986, to work as a summer intern at the Apple headquarters, he was taken on a tour to visit the Macintosh factory in nearby Fremont. Every day, truckloads of components and parts came in, just enough for one day's production; and every day, containers of Macintoshes were shipped out. The net result: no local storage, no warehousing. Vladimir was witnessing the very efficient technique of just-in-time (JIT) manufacturing.*

Ask anyone where background is located in a scientific paper, and the answer will consistently be 'the introduction'. Well, yes… and no. If background material is of no immediate use to the reader, it rapidly fades out of memory.

> ### The variable types
>
> *There are two types of variables in a computer program: global and local variables. Why? To allow the program to manage computer memory space more efficiently. Global variables require permanent memory storage whereas local variables free up their temporary memory storage space as soon as the program exits the subroutine where they were used. Could this wonderful concept apply to writing?*

Parking *all* background material in the introductory sections of your paper greatly increases the demands on memory. Background material comes in two flavors: the global background, applicable to the whole paper; and the local or just-in-time background, useful only within a section or paragraph of your paper. The just-in-time background imposes no memory load: it immediately precedes or follows what it clarifies. Here is a just-in-time example.

Additional information is readily available from "context" — other words found near the word considered.

In this example, the word "*context*" is defined as soon as it appears.

☀ When a heading or subheading in your paper contains a new word requiring explanation, explain it in the first sentence under the heading, in a just-in-time fashion.

Lysozyme solution preparation
Lysozyme, an enzyme contained in egg white, ...

In this subheading from an article for a chemistry journal, the word '*lysozyme*' is expected to be new to the reader. The writer defines it just-in-time in the first sentence of the section, using what grammarians call "an apposition" — a phrase that clarifies what precedes it. Kept short, appositions are very effective. Long, they are ineffective as the following sentence demonstrates.

Lysozyme, a substance capable of dissolving certain bacteria, and present in egg white and saliva but also tears where it breaks down the cell wall of germs, is used without purification.

Appositions are also ineffective when they slow down reading and lengthen a sentence as in the earlier example.

*The cellular automaton (CA) cell, **a natural candidate to model the electrical activity of a cell**, is an ideal component to use in the simulation of intercellular communications, **such as those occurring between cardiac cells**, and to model the abnormal asynchronous propagations, **such as ectopic beats**, initiated and propagated cell-to-cell, regardless of the complexity of their patterns.*

Select each custom-worded heading and subheading (i.e., not the standard *methodology*, *results* and *discussion* headings). Read the sentence that immediately follows that heading/subheading. It should provide the background for the heading. If not, modify that sentence.

The Broken Couple

> Wasted water, wasted thoughts
>
> *Vladimir stood still, hands under the hot water tap waiting for the water to become warm, wasting cold water down the sink. His problem-solving brain kicked into gear.*
>
> *"Why don't plumbers place the water heater close to the hot water tap, he wondered. Alternatively, I suppose one could use plastic pipes or insulate the heat-sinking metal pipes. Of course, I could always buy a thermostatic tap... with my next bonus check."*
>
> *His musings were interrupted by the arrival of the hot water and by the thought that he already had these thoughts before.*

When reading a sentence in which the verb never seems to arrive, has it occurred to you that your reader may also waste, or worse, be distracted by the words that separate the subject from its verb? Details inserted between the main components of a sentence **burden** (burden comes from the old French *bourdon*, a hum or buzz — but do we need to know that?) **the memory** because they move apart two words that the reader expects to see together, such as the verb (*burden*) and its object (*the memory*) in this sentence.

☀ Keep these happy couples close to each other.

• Verb and its object	• Subject and its verb	• Visual & its complete caption	• Background information and what requires background
• Unfamiliar word and its definition	• Acronym and its definition	• Noun/phrase and its pronoun	...

Separating the subject and the verb, as illustrated in ☞1, can be devastating.

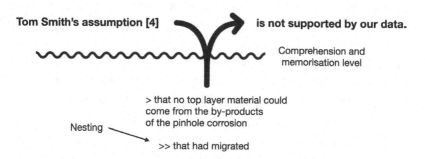

Figure ☞1
The nesting of phrases has the same effect as plunging the reader below comprehension level. In the end, all of the details below the comprehension level will be forgotten. Two broken couples are responsible for this: (1) The subject ('Tom Smith's assumption') is separated from the verb ('is not supported') by two nested sentences starting with 'that'; and (2) the phrase 'the by-products' is separated from 'that had migrated' by the phrase 'of the pinhole corrosion,' creating some confusion. It is not the corrosion that migrates, it is the by-products; To avoid the confusing double nesting, the writer could have changed 'that had migrated' into the noun 'migration' as 'that no top layer material could come from the migration of the pinhole corrosion by-products.' The sentence would have been clearer.

> ### Memory registers
>
> *Curiosity pushed Vladimir to open up his antique TRS80 computer to look at the printed circuit board. He could not quite recall what CPU was in use back then. Close to the heat sink, he spotted the long CPU chip, the Intel 8085 microprocessor. He remembered studying its structure back in 1981. That was when he discovered for the first time that rapid access to memory is so critical to the overall speed of a microprocessor, that the central processing unit (CPU) has its own dedicated memory registers right on the chip, under the same roof so to speak. This collocation allows ultra-fast storage and retrieval of data from these internal registers compared to the time it takes to store and retrieve data from external memory.*

☀ To increase reading performance through fast memory access, keep syntactically or semantically closely related items on the same page, in the same paragraph, in the same sentence, or on the same line. The reader will appreciate the increase in reading speed.

The Word Overflow

Memorizing takes a great deal of attention. The process is also slow. Have you ever been able to absorb complex road directions without asking the person to repeat them? Our working memory is not very elastic; it cannot accommodate a sudden word overflow as the next example demonstrates. Before you start reading it, a little background will help. This sentence is about molding machines. These machines melt plastic pellets and inject the molten plastic inside molds to make plastic parts.

> *"The main difference between the new micro molding machine design and the conventional « macro » molding machines with reciprocating screw injection system is that by separating melt plastication and melt injection, a small injection plunger a few millimeters in diameter can be used for melt injection to control metering accuracy, and at the same time a screw design that has sufficient channel depth to properly handle standard plastic pellets and yet provide required screw strength can be employed in micro molding machines."*[1]

Chances are you were not able to finish reading this 81-word sentence without having to read again parts of it. Its syntax is acceptable, and the meaning clear enough for a specialist familiar with the machine. But the working memory necessary to process it in one reading is too large for most readers.

When noticing such a long sentence, the writer is often tempted to break it down into two parts, by inserting a full stop in the middle of it.

> *"The main difference between the new micro molding machine design and the conventional « macro » molding machines with reciprocating screw injection system is that by separating melt plastication and melt injection, a small injection plunger a few millimeters in diameter can be used for melt injection **to control metering accuracy. At the same time** a screw design that has sufficient channel depth to properly handle standard plastic pellets and yet provide required screw strength can be employed in micro molding machines."*

Of course, it helps, but it rarely is the best way to deal with a long sentence.

---

[1] Zhao J, Mayes RH, Chen GE, Xie H, Chan PS. (2003) "Effects of process parameters on the micro molding process", *Polymer Eng Sci*, **43**(9): 1542–1554, ©Society of Plastics Engineers

☀ Before rewriting a long sentence to avoid what computer programmers call memory stack overflow and humans call headache, identify the intent of the author.

This sentence, found in the introduction of the paper, identifies a novel solution. The novelty is buried right in the middle of the long sentence in a place unlikely to attract much attention.

Before breaking down that sentence into multiple sentences, let's adopt a classic structure: start with the existing problem — then present a new solution. Unlike the original sentence that immediately plunges the reader straight into the core of the innovation, let us start with the known information — the background familiar to the reader.

> In conventional « macro » molding machines with reciprocating screw injection, melt plastication and melt injection are combined within the screw-barrel system. In the new micro molding machine, screw and injector are separated. The screw, now redesigned, has enough channel depth and strength to handle standard plastic pellets, but the separate injector, now with a plunger only a few millimeters in diameter, enables better control of metering accuracy.

The rewritten paragraph has three sentences instead of one, and 67 words instead of 81. The memory is not strained because the punctuation provides enough pauses for the brain to catch up: two commas and one full stop in the original versus five commas and three full stops in the rewritten version. The new version is also clearer and more concise.

Before we leave this example, and since good writing is also convincing writing, I would like to emphasize something the authors of the original sentence did well. They anticipated that the users of the new 'Micro-Molding' machines would want to use the same (cheap) plastic pellets that they use in their old 'Macro-Molding' machines. They may have tons of these pellets stored on wooden pallets in their warehouses. The authors responded to a possible objection by reassuring the users that the redesigned screw is able to handle standard plastic pellets.

☀ The best way to persuade someone to let go of something familiar and adopt something new, is to anticipate and respond to the most common objection people might have when considering the change.

Choose a section of you paper (start with the introduction if you want). Select the sentences that look long. Look at the number of words in that sentence (bottom line of the window). If that sentence exceeds 40 words, read it to see if it is clear. When unsure, revise for clarity: identify the purpose of the sentence and look for a classic structure to help you break it down into smaller sentences. Alternatively, you may want to remove the excessive details that make it long, or rewrite the whole paragraph.

In summary, acronyms, pronouns, synonyms, abusive detailing, background ghettos, cryptic captions, disconnected phrases and long sentences, all take their toll on the reader's memory.

Chapter 5

Sustain Attention to Ensure Continuous Reading

Attention-getters

Vladimir and his wife, Ruslana, usually read in bed before going to sleep.

"You haven't finished it yet?" Ruslana asked.

It was more a remark than a question. Her husband had been reading a 10-page scientific article for the past three nights, while in the same amount of time, she had read close to 250 pages of a suspenseful novel. Each evening, she had remained in bed; whereas Vladimir, unable to remain focused, had been in and out of bed, for a drink, a phone call, the late-night TV news, or a snack. She knew the signs. Tired after a long day at the lab, he did not have enough energy to stay attentive for more than 10 minutes at a time. His article required too much time and concentration. She asked Vladimir,

"Have you ever read a really interesting scientific article that you could not stop reading?"

He looked at her, and remained silent long enough for her to know that there could not have been many.

"I can't think of one!" he said finally, "Even my own papers bore me."

"Don't they teach you how to write interesting papers at your Institute, you know, papers that attract the attention of scientists?" Vladimir sighed.

"Attracting is fine. It is sustaining the attention that is hard. I wish I could keep my readers as awake and interested as you are."

She could not resist the tease.

"Oh, I am awake, darling, and interested. Are you?"

Love, drama, and suspense add spice to novels. But how can there be any suspense in a scientific paper when the writer reveals all secrets right upfront in the title? Imagine crime author Agatha Christie writing a novel with a title like "The butler killed the Duchess with a candlestick in the library". The book would surely make it to the top of the worst-seller list — at least, as far as suspense goes!

It would start with a summary that says "In this book, we will show that it was the butler who killed the Duchess with a candlestick in the library." The introductory chapter would mention that the chamberlain, the gardener, and the maid also live with the Duchess, but none of them have any motivation to kill her. The middle chapter would show the photo of the butler caught in the act of bashing the Duchess over the head with a candlestick. Additional chapters would include the signed testimony of the butler admitting to the murder. And the final chapter would reinforce the facts already established: the weight of the candlestick and the muscular strength of the butler are enough to kill the Duchess. It would end with a proposal for further research, such as the criminal use of candlesticks by Geishas on Sumo wrestlers. You have here a possible rendition of a scientific paper à la Agatha Christie, but not by Agatha Christie.

How would Agatha Christie make any scientific paper more interesting? Let me suggest that, for one thing, she would turn your contribution into more of a story. She would bring twists into your story plot, mention unexpected problems, which you, the scientist hero deploying inductive and deductive logic, would victoriously solve. Sometimes, she would have you pause to elaborate or clarify, instead of blissfully moving athletically through your complex story with total disregard towards the breathless non-expert reader. And even though she could not use suspense at the top-level of the story to hook the reader, she would recreate local suspense, whenever the opportunity arose. There would rarely be a dull moment. Indeed, having listened to some riveting Nobel Laureate lectures, I am now convinced that a scientific story need not be boring at all.

Keep the Story Moving Forward

The most challenging word of this heading is the word 'story'. Somehow, scientists do not see the active story as an appropriate model for a scientific paper. Granted, not all parts of a scientific paper gain clarity and garner interest through the story style (the methodology section rarely does), but all parts can benefit from the techniques presented next.

☀ The great attention-getter, **change**, is what keeps a story moving forward.

Take a change in paragraph, for example. With each change, the story progresses, widens, narrows, or jumps. When ideas stop moving, a paragraph lengthens. Imagine a river. When does water stop moving?

Sometimes, the river widens to form a lake with no discernible current. Ideas stop moving when they stagnate or expand into paraphrases. Sometimes water gets caught in a whirlpool. The author traps the reader in a whirlpool of details before returning to the main idea. Sometimes the river has deep meanders. The author makes an unexpected U-turn to complete an already presented idea. Sometimes slow counter-currents form alongside the riverbank. The author, inverting subject and object positions in a sentence, slows down reading because of constant use of the passive voice — a proven story-killer.

The lake

When there is no purpose (because there is no current), a paragraph grows in length. Ideas are presented in no specific order, or in an order obvious only to the writer. Extensive paraphrasing may also take place, unnecessary lengthening the paragraph until it resembles a hefty chunk of text that is discouraging to the eye. Without having to read a single word, the reader knows by experience that reading will be slow, and clarity will be low.

> When ideas are not in motion, a paragraph grows in length. The additional length slows down reading and reduces overall conciseness. **With paraphrasing, the paragraph lengthens without actually moving the ideas forward since the sentences have the same meaning. ...**

The sentence in bold paraphrases what the first two sentences already cover.

The whirlpool

Nested detailing (when details explain details) also prolongs paragraphs. Nested detailing takes the reader away from the main intent of the paragraph. When the details stop, the reader is ejected back into the main topic stream. In the next paragraph on embryonic cell proliferation in culture dishes, the in-depth description of the culture dish (text in bold) distracts the reader (dish → coating → reason for coating). These details could have been described elsewhere, or simply removed.

> For the next three days, the thirty embryonic cells proliferate in the culture dish. **The dish, made of plastic, has its inner surface coated with mouse cells that through treatment have lost the ability to divide, but not their ability to provide nutrients. The reason for such a special coating is to provide an adhesive surface for the embryonic cells.** After proliferation, the embryonic cells are collected and put into new culture dishes, a process called 'replating'. After 180 such replatings, millions of normal and still undifferentiated embryonic cells are available. They are then frozen and stored.

The omega meander

The reader is distracted when the author makes an unexpected Ω turn to add detail to a point made several sentences earlier. The flow of thoughts is disrupted. In the following example, the sentence in bold should follow the first sentence to remove the meander and linearize the flow.

> After conducting microbiological studies on the cockroaches collected in the university dormitories, we found that their guts carried microbes such as staphylococcus and coliform bacteria dangerous when found outside of the intestinal tract. Since cockroaches regurgitate food, their vomitus contaminates their body. Therefore, the same microbes, plus molds and yeasts are found on the surface of their hairy legs, antennae, and wings. **It is not astonishing to find such microbes in their guts as they are also present in the human and animal feces on which cockroaches feed.**

The counter current

A normal sentence has current. It pulls reading forward from the old information upstream to the new information downstream, from the sentence' subject to its verb and object. When such a natural order is unduly inverted (with the passive voice for example), the sentence commonly lengthens, and reading slows down.

The cropping process should preserve all critical points. Images of the same size should also be produced by the cropping.

Look at your long paragraphs and ask yourself, am I making a single point here? Can I make that point using fewer arguments, fewer words, or even a figure? Would making two paragraphs out of this one paragraph clarify things and keep ideas in motion? Do I have lakes, whirlpools, meanders, or counter currents?

Twist and Shout

> ... Shout
>
> *'Twist and Shout,' a song The Beatles made popular, is an attention grabbing heading that requires explanation. The twists, in our attention-grabbing context, are the twists in the story plot. The shouts are all things that call the reader to attention.*
> *Shouting is quite easy!*

☀ Turn up the visual volume; add subheadings and contributive visuals.

Surrounded by vision-enhancing white space, subheadings shout in their bold font suits. Subheadings tell the reader how your story is moving forward. Therefore, make your subheadings as informative and indicative of content as possible. Avoid vague subheadings such as Simulation or Experiment, but also avoid excessively specific ones that contain keywords not even in your abstract.

☀ Bring changes to format and style.

Used in moderation,

1) A numbered list,
2) **Bold**
3) <u>Underlined</u>
4) *Italic text*

5) **a change in font**

6) A box around text

are equivalent to raising the volume of your voice, or changing its pitch or prosody. These changes break the monotony of paragraphs and make things stand out (note that the publisher may limit your choices by imposing a standard format and regulating the use of styles).

☀ Change sentence length, and ask questions.

After a long sentence, and particularly at the end of a paragraph, a short sentence carries much emphasis. It shouts. Why? It does not clog the memory, its syntax is fast to process, and its meaning is easier to understand. The last sentence of the following paragraph is four times shorter than the three sentences that precede it. That short sentence not only shouts, it screams!

> *Annotations on paper photos were all manual; they were either implicit (one photo album by event, location, or subject) or explicit (scribbles on the back of photos). In today's digital world, while some annotations are still manual, most, like time, date and sometimes GPS location, are automatically entered in the photo file by the camera. Is it conceivable that one day automatic annotations will be extended to include major life events, familiar scenery, or familiar faces, thus removing the need for manual annotations? Yes, and that day is upon us.*

The theme of the paragraph is photo annotation — yesterday, today and tomorrow. The passive voice used throughout the paragraph is just a writer's tool to achieve the writer's purpose: automation is inevitable. You may argue that it is possible to rewrite this paragraph using the active voice. I agree, and to show you that the active voice is not a panacea, healing all writing problems, here is the "active" paragraph. This example will also help us discover one more attention sustaining technique.

☀ Change sentence syntax and length.

> *Annotating paper photos was a manual task, implicit (photo albums by event, location, or subject) or explicit (scribbles on the back of photos). Annotating digital photos is an automatic task for digital cameras (time, date, and sometimes GPS location inserted in the photo's data file). Annotating automatically the previously manually entered data will*

tomorrow be possible for major life events, familiar scenery, or familiar faces.

All three sentences share a constant topic located upfront ('*annotating*'). All three sentences are written in the active voice. All three sentences have similar lengths. And all three sentences follow exactly the same syntax: subject, verb, and object. Parallelism in length and syntax over more than two sentences loses its magic. The reader is bored. A statement now replaces the suspenseful question. And even though the paragraph is more concise (65 words versus 88), it is now lifeless.

☀ Convey importance with words that act like pointing fingers.

Certain words excel at guiding the attention if used sparingly. if overused, they lose all meaning, and in fact can begin to annoy the reader. These words give you a way to reveal the salient facts amongst all the facts.

> *more importantly, significantly, notably, in particular, particularly, especially, even, nevertheless*

☀ Keep the same contribution in front of the reader throughout your paper.

To make sure that the reader never loses sight of the paper's contribution, the writer scientist mentions it in every part of the paper: the title, the abstract, the introduction, the conclusion, the figures, the body of the paper, and even title-echoing words from the headings and subheadings. But if the writer cloaks the contribution with synonyms, if the writer seems to tell a different story in the title, abstract or conclusion, or if the writer digresses, then the reader may wonder what the contribution really is.

... Twist

There are always opportunities for tension in a scientific paper. It may be a limit you are about to circumvent, or an exception you are ready to exploit. It may be a common point of view you are about to change, or a gap you are about to bridge.

☀ Announce contrasted views or facts.

Special introductory words will bring you as close to drama as you'll ever be in a scientific paper.

> *however, but, contrary to, although, in contrast, on the other hand, while, whereas, whilst, only, unable to, worse, the problem is that …*

Things are not always as perfect as they look at first sight.

> *"**Although** COBRA (Cost Based operator Rate Adaptation) has shown itself to be beneficial for timetabling problems, Tuson & Ross [266, 271] found it provided **only** equal or **worse** solution quality over a wide range of other test problems, compared with carefully chosen fixed operator probabilities."*[1]

Twists are tales of the unexpected. Too often scientific papers are tales of the expected. If papers are dull, it is because all difficulties encountered during the research have been erased from the public record. Only what works is presented.

☀ Keep enough of the unexpected difficulties to sustain interest and
build in the reader's mind the image of a resourceful scientist.

> *interestingly, curiously, surprisingly, might have (but did not), unexpectedly, unforeseen, seemingly, unusual, different from, …*

In the next example, the modal verb '*might have*' intrigues the reader. … *might have, but did not!* It sets the expectation that the writer will explain why the method is not as applicable as originally thought.

> *The Global Induction Rule method [3], a natural language processing method, **might have** worked on news video segmentation since news contents can be expressed in a form similar to that used for text documents: word, phrase, and sentence.*

☀ Use numbers.

Too many numbers distract the reader and clog the memory, but a couple of well-chosen numbers grab the attention. Numbers have considerable attractive power.

---

[1] Reprinted with permission from Dr. Mark Sinclair, PH.D thesis, "Evolutionary Algorithms for optical network design: a genetic-algorithm/heuristic hybrid approach", ©2001.

After 60 years old, your brain loses 0.5% of its volume per year.

0.5% makes this sentence precise and dramatic (at least for the elderly), and *'millions'* attracts attention in the following example.

*After proliferation, the embryonic cells are collected and put into new culture dishes, a process called 'replating'. After **180** such replatings, **millions** of normal and still undifferentiated embryonic cells are available.*

In the title and the abstract, you do not have the artillery power of visuals to convince the reader that your paper is worth reading. The next best thing to help you convince the readers that they really need to download your paper and use your findings is... NUMBERS. Is your abstract precise or vague? Would numbers be appropriate in your title? Inside your paper, you have visuals. Some journals propose a ratio of one visual per thousand words. Calculate the ratio between the number of columns taken by visuals (include the visual text captions as part of the space occupied by visuals) and the number of columns taken by paragraph text (exclude the title, abstract and reference parts of your paper). Now look at the most highly cited papers mentioned in the references at the back of your paper, and calculate the same ratio (visual versus text). How does your ratio compare to their average ratio? Do you have enough visuals, i.e., are you sufficiently convincing and clear?

☀ Announce the alternative routes when you are about to change direction in your story plot.

rather than, instead of, alternatively

*"**Instead of** unidirectional motion along a single pathway, can we have unguided motion through the myriad of shapes?[2]"*

Pause to Illustrate and Clarify

After a particularly long or arduous section of any scientific paper, the reader's ability to absorb more knowledge is low. A sign that this is happening is the inability to recall much of what has been read (if it had been clear, memory recall would have been much easier).

---

[2] Reprinted excerpt with permission from Wolynes PG. (2001) Landscapes, Funnels, Glasses, and Folding: From Metaphor to Software. *Proceedings of the American Philosophical Society* **145**: 555–563.

Another sign is the inability to distinguish the important from the less important (dependencies, causalities, and relationships are not yet fully identified). These low points in knowledge acquisition happen time and time again while reading the paper. At these times, reader attention is easily lost when it ought to be sustained. Maybe it is caused by too much specialized vocabulary or dense theoretical considerations like formulas. The writer should identify such places in the paper, and plan for a pause to clarify and consolidate. To clarify, the writer uses illustrations, examples and visuals. To consolidate, the writer uses brief summaries. Clarifications and consolidations are highly appreciated by the reader, and temporarily boost the reader's attention level.

☀ Summarize (be it in text form or visual form) to clarify what is important.

A summary rephrases succinctly and differently the main points. It gives readers a second chance to understand, and it gives writers the assurance that readers will be able to keep in step with them. The following words perk up the attention of readers because, in them, they see the promise of knowledge consolidation.

To summarize, in summary, in other words, see Fig. X, in conclusion, in short, briefly put, …

Reading is hard, but writing is harder. Distilling years of research in less than 10 pages is a perilous exercise. Like compressed audio files, compressed knowledge loses clarity. Even if the structure of your paper is clear, you need to reintroduce detail into your text to make things clear, easier to grasp, less theoretical, more practical, and more visual.

☀ Use examples to illustrate.

The need for **examples** is not just a by-product of the distillation process. Illustrative details are needed because, more often than not, your readers are not familiar with what is happening in your field of research. They may be scientists in the same domain, but the distance between you and them in terms of knowledge is great, regardless of their academic level. What is tangible and real to you, may just be an idea, or a theory to them.

Your concern for making things clear to the reader is shown through words and punctuation. The words *for example, namely, such as, in particular, specifically,* and the colon, all keep the attention of the reader elevated because they promise easier understanding, less generalities, and more clarifying details. They bring relief to struggling readers, and wipe out deep concentration furrows in their brow.

These words also reveal your expertise. Non-experts cannot give examples or specific details. Their comfort zone is in the general and the imprecise. Experts, on the other hand, are at ease in the specific, the precise, and the detailed. That is what makes them convincing, believable, and interesting!

 Take your paper. Search for the words *summarize, summary, conclude,* and *in short.* If every time the search window returns empty, ask yourself: Are there sections in my paper where I can anticipate that the reader will be unable to see the tree for the forest, to identify what is really central in my argument or findings? If your answer is yes, then end these sections with a brief summary.

Recreate Local Suspense

The structure of a scientific article leaves little room for suspense. The gist of the contribution is revealed immediately in title and abstract, well before the reader reaches the conclusions. Therefore, suspense has to be artificially recreated.

Do you know what might reduce the accelerated shrinking of your brain and possibly slow down the onset of dementia in your old age?

☀ The master of suspense, the most underused and underrated tool in the writer's toolbox is... the **question** mark '?'

It is the quintessential universal trigger to guided thought processes. Questions make people curious. As soon as the writer expresses a question, it becomes the reader's question and remains so until the answer comes.

(a) A question refocuses the mind and prepares it for an answer.

(b) A question establishes the topic of a paragraph clearly.

(c) A question moves ideas in a given direction.

(d) A question lingers in the mind until it is answered – oh, by the way, the answer to the shrinking brain question is *Vitamin B*.

Does one always need a question mark to ask a question?

No. The question mark makes a question explicit, but there are ways of raising implicit questions. In grammar, the only question that comes without a question mark is **the indirect question**. It is asked by the author.

We wondered whether our data cleaning method was valid.

Unexpectedly, might have, surprisingly, and *curiously* announce **unexpected findings** that raised questions first in the mind of the author and later, by proxy, in the mind of the reader. In the next sentence the intriguing *would appear* raises a question. Manual polishing only appears to be the answer, but clearly, the writer has a better, albeit less obvious answer.

*What method provides enough contact force to polish the highly complex surfaces of large ship propeller blades? Manual polishing with a belt machine **would appear** to be the obvious answer.*

The **not-yet-justified adjectival claim** always raises a question, particularly when it is emphasized as in the following sentence:

*"The energy landscape/funnel metaphor leads to a **very different** picture of the folding process than the pathway metaphor.[3]*

In this case, the question is "What does the picture of the folding process look like when we use the energy landscape/funnel metaphor?" The author prompts the reader to ask that question by making the claim before bringing in the evidence. What a marvelous way to recreate suspense!

*At high temperatures, electron trapping is **not** ...*
Electron trapping is unimportant when ...

Acting in a manner similar to the adjectival claim, is the **negative statement** that presents what does not work before what works, what is unimportant before what is important. Starting a statement with what doesn't, begs the question of what does!

---

[3] Reprinted excerpt with permission from Wolynes PG. (2001) Landscapes, Funnels, Glasses, and Folding: From Metaphor to Software. *Proceedings of the American Philosophical Society* 145: 555–563.

The **announcement of change** increases tension and attention.

The new strain of Malaria had become resistant to all existing medicine, but this was about to change.

The **provocative statement** is a trumpet call for justification or clarification. The following statement is as provocative as the title of Thomas Friedman's book: "The World Is Flat."

HTML 5 is stillborn.

Certain **values in a visual** raise strong questions. Yes indeed, what happened on Thursday?

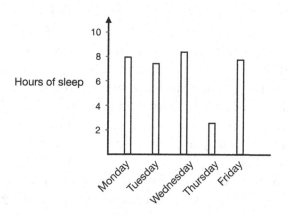

Antagonistic claims create natural suspense. Observe how the author sustains the interest. In five consecutive sentences, mostly written in the active voice, he brings (1) two numbers, (2) a figure, (3) the attention-getters *'but'*, *'not'*, and *'important'*, (4) an antagonistic claim *'whereas'*, *'contradiction'*, and (5) a question suggesting one of the reasons for the observed difference in results.

> *When trying to improve the clusters obtained by the Clusdex method, Strunfbach et al. reported a <u>27%</u> increase in error rate when using our annealing method [6], whereas we had observed, not an increase, <u>but</u> a <u>12%</u> decrease in error <u>as seen Fig. 3</u>. Their findings represented <u>an important contradiction</u>. After examining their data, we discovered that they had used the raw data while we had removed the outliers. Should outliers be kept or removed?*

Roadblock. Different methods, different results, and no way to compare them create an impasse.

"As was pointed out (3), it is a challenging task to compare the results of these profiling studies because they used different microarray platforms that were only partially overlapping in gene composition. Notably, the Affymetrix arrays lacked many of the genes on the lymphochip microarrays ..."[4]

Search your entire paper for the question mark. If you do not have any, ask yourself why? What are the essential questions you are raising and answering in your paper? Shouldn't you find a way to express them in question form? Do you recreate suspense and raise questions by other means such as roadblocks, provocation, visual questions, negative or adjectival claims?

At the beginning of this chapter, we identified that the structure of a scientific article leaves little room for suspense. The gist of the contribution is revealed immediately in title and abstract, well before the reader reaches the conclusions. Therefore, you, the author, have to reintroduce tension and suspense to sustain interest. Yes, it may feel unnatural; it may also add length to your paper. But the benefits far outweigh the costs. Because attention-getting schemes tend to increase contrast, clarity also increases. More attentive, the reader gets more out of your paper.

In this chapter, we also examined many ways of sustaining the initial attention of motivated readers as they plow through the tough parts of your paper where their attention drains at a fast pace. We used words, punctuation, syntax, text style, page layout, structure, examples, summaries, visuals, and questions. Since the attention of your reader is not deserved but earned through hard work, it is time for you to practice. It is time to wake up the comatose reader with adrenaline-loaded questions. It is also time to pour life-restoring liquid details onto desiccated knowledge.

Read your paper to identify the parts the reader might find hard to understand (give it to somebody else to read if you are unsure). In these parts, modify your text accordingly to increase attention and facilitate understanding (examples, visual aids, questions, new subheadings, etc.).

---

[4] Wright G, Tan B, Rosenwald A, Hurt E, Wiestner A, Staudt LM. A gene expression-based method to diagnose clinically distinct subgroups of diffuse large B Cell Lymphoma. *Proceedings of the National Academy of Sciences*, **100**: 9991–9996.

Disclaimer: Attention getters are only effective if used sparingly. The author of this book will not be responsible if excessive use of attention-getters in a paper distracts readers and makes them lose focus. Examples of such excessive use include writing 'importantly' seven times in a long section, turning a paper into a cartoon through over-abundant visuals, using bold and italic in so many places that the paper looks like a primary school paper, or starting three consecutive sentences with 'however'.

Chapter 6

Reduce Reading Time

Reading time is reduced when the writer is concise and to the point. Conciseness is seen by some authors as a mark of respect for the reader, as Pascal's quote illustrates.

> Blaise pascal
>
> *The 17th century French mathematician and philosopher Blaise Pascal apologizes to his reader in his book, lettres provinciales. He writes: "This letter is longer, only because I have not had the leisure of making it shorter." Conciseness seen as politeness! Boileau, a French writer from the same century, has harsh words: "Who knows not how to set limits for himself, never knew how to write."*

Although time is measured in seconds, reading time is much more subjective. Readers experience the passage of time differently for different reasons: familiarity with the topic, linguistic skills, reading motivation, reading habits, or the reader's tiredness. Using techniques, the writer is able to reduce both objective and subjective time while increasing overall reading pleasure.

Readers are less sensitive to time when they look at visuals. Somehow, the brain is more engaged looking at a visual than reading paragraphs of text. Good visuals are information burgers devoured like fast food. The eyes rapidly scan them, detecting patterns as they go. Their message reaches the brain faster because processing them takes place in the highly developed visual cortex designed to handle **high information bandwidth**. For text or speech, the brain sequentially processes one word or syllable at a time — a low information bandwidth process. Should you need convincing, conduct the following experiment. Which of the following two representations gives you the largest amount of information in the shortest amount of time: the following text

In our experiment, for an up flow velocity of 0.10 m/h, the observed normalized tracer concentration of the effluent increased rapidly from 0 to 0.4 after 15 hours. The increase slowed after 38 hours when the concentration reached 0.95. It peaked at 1.0 at 90 hours. Following which, the concentration curve decreased steeply down to reach zero asymptotically at 180 hours. The calculated data and the observed data were closely related. However, when compared with the calculated data, the observed data seemed to lag when the concentration dropped.

or figure 🖙 1?

When reading is slow, readers become impatient. So examining what makes reading slow is essential if the writer wants to give the reader a pleasurable reading experience.

Reading is slow when the writer creates reading accidents, such as undefined acronyms, ambiguous pronouns, long sentences, missing logical steps, or unfilled knowledge gaps.

Reading is slow when the sentence is too complex or abstract. Pascal wrote that *"Memory is essential for the operations of reason".*

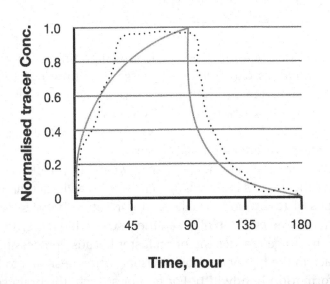

Figure 🖙 1
Tracer concentrations (dotted line) for an effluent velocity of 0.10 m/h. Calculated data (solid line) lag experimental data when concentration drops.

Complexity makes great demands on anyone's memory. It considerably slows down reading.

Reading is slow when the structure is insufficient. Readers spend more time when they cannot easily find what is of interest to them for lack of a clear structure with enough headings and subheadings.

☀ Make sure you have a detailed, informative structure (more on that later).

Structural elements are quick to read, short and simple in syntax, easily understood by the reader, and easy to locate and identify. For example, the following sentence in bold receives great attention because the indentation creates a white space acting as an eye-trap.

Reading is slow when the reader is not as familiar with the topic as the expert is. For the people new to your field, paradoxically, a longer introduction reduces the overall time required to read your paper simply because it sets the foundations needed to understand the rest of the paper. Veteran writers, after writing their paper to achieve maximum clarity and conciseness to satisfy experts, spend more time on their final drafts to cater for the needs of the non-experts. Here is what UC Davis Professor Dawn Sumner has to say about her writing process:

> "I leave [my latest draft] for a while, and then I think about what my audience already knows, and I give them all the information that they need [so that] the flow follows for someone not as familiar with the topic as I am, and then [I] sort of reiterate that."

Let me argue that anyone's writing will never be excellent until, like Professor Sumner, the writer keeps the reader in mind. Thinking about what the reader may not know allows you to identify the knowledge gaps that will slow down reading, because the reader has to work hard to understand how your sentences are logically linked.

☀ Sentences that fill the knowledge gaps, and explanations of gap-creating keywords always speed up reading.

Reading is slow when extra processing time is required because the syntax is complex. For example, readers, experts or non-experts, struggle with long compound nouns. These require extra brain cycles to decode, which slows down reading.

☀ Unpack long compound nouns and clarify them by adding a preposition (of, on, to, with) to speed up reading.

Reading is slow when writing lacks conciseness. As Pascal points out, a lengthy paper takes less time to write than a short one, but it takes more time to read. Identifying the sources of excess length at a global level is the first step towards conciseness, but it is as difficult as determining the causes of a bulging stomach. The need for the diet is clear, but the fat can come from so many sources. Where is the fat in your writing?

- Length is caused by the thousand words that should have been a diagram, a graph, or a table.
- Length grows out of a structure still in the formative stage, where information is needlessly repeated in different sections.
- Length is born out of the slowness of the mind, as it warms up and spreads a fog of platitudes, particularly in the first paragraph following a heading or subheading.
- Length is the fruit of unrealistic writer ambitions, aiming at cramming in a single paper the contribution of several papers.
- Length is the fruit of hurriedness, since it takes time to revise a paper for conciseness.
- Length happens when the reader is given details of the unnecessary kind, details that do not enhance the contribution.

Can your writing be too concise? Yes, it can, and your text would then lose clarity. Here are four good reasons to justify lengthening a paper.

- Lengthen to write a longer introduction that really sets the context and highlights the value of your contribution. Your contribution is like a diamond. To hold and display it, you need a ring (the ring of related works).

- Lengthen to highlight aspects of your contribution in every section of the paper (different angle and level of detail). Each facet of a diamond contributes to its sparkle. Likewise, each part of a paper presents the same contribution at a different angle.

- Lengthen to go beyond stating results in the abstract, and reveal the potential impact of your contribution on science. Would you give an uncut diamond and ask readers to polish it themselves?

- Lengthen to provide the level of detail that enables research colleagues to independently assess your results or at least follow your logic.

Read your paper. Are you repeating details? If you are, revise the structure to avoid repetition. Do you feel that readers of the journal in which you publish your paper already know what the first paragraph of your introduction says? If you do, cut it out. Is the last paragraph of your introduction a table of contents for the rest of the paper? If it is, cut it out. Are you bored reading your own prose? If you are, it is time to replace it with a visual. Are all details essential to your contribution? Read the whole paper again and cut ruthlessly the details that explain details.

Chapter 7

Keep the Reader Motivated

Attention is a precious resource, not to be wasted. It is the traffic controller in your brain, prioritizing whichever thought it deems most important. If attention wanes, our train of thoughts could derail or be redirected away from reading. Yet, for all its importance, attention is governed by a powerful ruler: motivation. Consider reading as a system with inputs and outputs, as shown in ☛1.

Motivation is one of the four critical inputs to the system. The fun of gaining knowledge (feedback loop) keeps motivation high or even increases it (for example when expectations are exceeded or when goals are met quickly). Motivation decreases when expectations are not met (syntax is too obscure, initial knowledge is insufficient, or paper is disappointing), when alternatives to reading become more attractive, or when the reader is tired.

> ### Wrong title and unmet expectations
>
> *Vladimir is a young US researcher new to the field of English speech recognition. When Popov, his supervisor, sent him two months ago to the ICASSP conference in Paris to catch up with the latest happenings in the field, he could not have foreseen that this trip to France was about to put Vladimir in some sort of trouble.*
>
> *Today, young Vladimir is searching online journal databases for general articles on automated speech recognition in over-the-phone plane reservation systems using dialogues. Fifth in a long list of titles, he spots the title: "Over-the-phone dialogue systems for travel information access." He smiles. All the keywords he typed are there. He orders the paper through Joan, the research center's librarian.*
>
> *A day later, the paper lands on his desk with a yellow post-it note attached to the first page that says, "French girl friend you met in romantic Paris?"*

> *Vladimir does not understand. He is happily married to Ruslana, a Russian. Puzzled, he removes the post-it that covers the abstract and starts reading. The abstract is at odds with the title. He had hoped for a general article, but he finds that it is about French speech in dialogue systems. His eyes move to the name of the first author: Michelle Mabel. A French woman! Darn! No wonder the librarian is teasing him. Why else would he be reading an article so foreign to his research field? Should he start reading? Or should he worry about the rumor that is probably going around the lab about a torrid extra-marital affair with a French woman? ... (To be continued)*

Dash or Fuel the Hopes of Your Readers, Your Choice

Motivating the reader starts with the title of your paper. It provides the initial reading impetus. The reader will scan hundreds of titles and select only a few. Imagine for a moment that the reader found your title interesting. You have what all authors dream of: the reader's attention. So it is now up to you. Are you going to dash your reader's hopes, or on the contrary, fuel them?

Dash — by a title that is not representative of the rest of the article
Fuel — by a title that is representative of the rest of the article

> *Vladimir decides to read it anyway. After all, the paper is only five pages long. He should be able to get through it fast. He will just skip the parts that do not interest him.*
>
> *Half an hour later, he is only in the middle of page 2 reaching the end of the introduction, gasping for a graphic or a diagram to make things clearer. He glances anxiously at his watch. He has a meeting coming up with his team in 20 minutes. (To be continued)*

Dash — by making clear that reading the article will require more time than anticipated
Fuel — by making clear that reading the article will require less time than anticipated

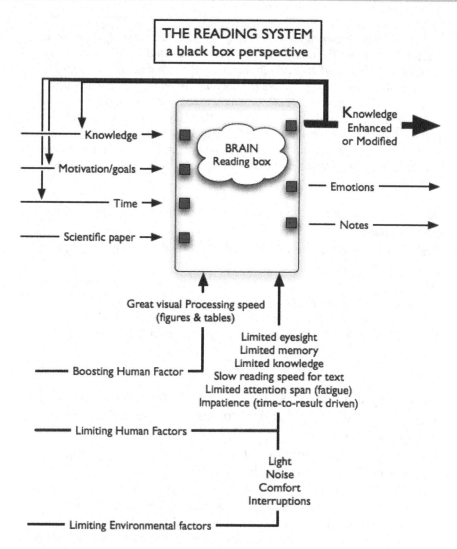

Figure ☞ 1

Reading, considered as a system, has four inputs and three main outputs. Prior to reading, each input has an initial value. That value will change over time because the outputs influence the inputs. For example, the more knowledge you get from the paper, the more knowledge you put back to facilitate further understanding. External factors influence the reading process. They either lubricate the process or create friction and inefficiencies. They indirectly affect the pace of absorption of knowledge, and therefore motivation, a critical input to the system. If reading was a transistor, motivation would be its base current that either shuts down or promotes the reading activity.

> *The introduction mentions French phonetics, and the differences in accents between the Chtimi and the Marseillais dialects. His knowledge of France is limited to football players and perfumes. He has heard of Zinedine Zidane, and Chanel #5 (he bought it in Paris for Ruslana's birthday), but that's about it. He glances at the reference section. No help there. He looks up Wikipedia online. No help there either. If he asks Joan, the librarian, she will ask him about his French girl friend, so he scraps that idea. By now, his motivation is at its lowest point, so he skips a few paragraphs and jumps to the part about dialogue modeling. (To be continued)*

Dash — by not giving the reader the baseline knowledge required to read the paper
Fuel — by providing the reader with the baseline knowledge needed to read the paper

> *However, he did not know that his motivation could drop lower than its previous low level. The key paragraph that seemed to be precisely in his area of interest is totally obscure. He spends a good 5 minutes on it, and then gives up. (To be continued)*

Dash — by using prose so obscure or complex in syntax that the reader gets discouraged and becomes unsure whether he or she understands correctly
Fuel — by using prose so clear that the reader is encouraged and sure that he or she understands correctly

> *He finally decides that the article is too specific. The semantic modeling will not apply to English at all. He won't be able to use it. And his meeting starts in a few minutes. A colleague who is also going to the meeting appears in his cubicle, looks quickly at one of the subheadings of the paper and says, "Hey, Vlad! I didn't know you were interested in French. Got a French girl friend? Does Ruslana know?"… Vladimir drops the paper in the trashcan. "It's a mistake," he says. (To be continued)*

Dash — by making the reader doubt the validity, or applicability of the contribution
Fuel — by demonstrating the quality, validity, or applicability of the contribution

> *He rushes to his meeting. As he enters the room, all his colleagues shout what sounds like 'Bonjour'. Before he answers, he looks around and relaxes; his boss has not yet arrived. He says, "It is not what you think. The title was misleading." They all laugh. At that moment, his boss enters the meeting room and hands Vladimir a paper: "Here, read this," he says. "I have not read it, but it seems perfect for your research. It's by... um, a French author." The whole group collapses in laughter.*

Dash — by boring the reader with a style lacking dynamism, a sentence structure lacking variety, new information lacking emphasis, and text lacking illustrations

Fuel — by captivating the reader with a dynamic style, a varied sentence structure, new information properly emphasized, and text rich in stories and illustrations

I deliberately chose a story to illustrate the various disappointments researchers experience while reading certain papers. How much more dynamic is the language of stories compared to the stiff classic scientific writing style. Deviating from the norm is often frowned upon (like ending sentences with a preposition). Yet, one section of your paper is ideally suited to accommodate such deviations: the introduction. You have a story to tell: the story of why you embarked on your research, why you chose a particular method, etc (see chapter on the introduction). Since it is a story, use the narrative story style to write it.

You can now see that your writing sustains the motivation of the reader through a combination of writing style, honest title, judicious detail and background, clear contribution, and good English.

Meet the Goals of Your Readers to Motivate Them

In our story, Vladimir (a newcomer to the field) is interested in general background. There are many kinds of readers, all coming to your paper with different motives and different levels of expertise. Satisfying and motivating them all is an impossible exercise if you do not really understand what readers hope to find in your paper. The following scenarios will help you understand their goals.

The field intelligence gatherer

Hi! I am a scientist working in the same field. I may not be doing exactly the same research, but I am a regular reader of the journal you read and attend the conferences you attend. I was the guy sitting on the fifth row facing you when you presented your paper in Korea last year. I have read most of your abstracts to keep up to date with what you're doing.

First among the six reader profiles examined is the intelligence gatherer. Such scientists are interested in anything in abridged form: your abstract or conclusions, sometimes the introduction. They probably will not read your whole paper.

The competitor

Hi! You know me, and I know you. Although we have never met face to face, we reference each other in our papers. By the way, thanks for the citation. I am trying to find a niche where you are not playing, or maybe I'll fix some of your problems in my next paper. Hey, who knows, maybe you are onto something I could benefit from. I'd love to chat or work with you on a common paper one of these days. Interested?

Even if some of your background is missing, competitors are able to fill in the blanks without your help. They will read your paper rapidly, starting with the reference section to see if their name is in it and if their own reading is up-to-date. They may also use your list of references to complete their own list. They will probably skip your introduction. Occasionally, competitors may be asked to review your article before publication.

The seeker of a problem to solve

Hi! You don't know me. I am a senior researcher. I just completed a major project, and I am looking for something new to do. I am not quite familiar with your field, but it looks interesting, and it seems as though I could apply some of my skills and methods to your problems and get better results than you. I am reading your paper to find out.

Problem seekers may read the discussion, conclusions, and future work sections of your paper. Since their knowledge is not extensive, they will also read the introduction to bridge their knowledge gap.

The solution seeker

Help! I'm stuck. My results are average. I am pressured to find a better solution. I need to look at other ways of solving my problem. I started looking outside my own technology field to see if I could get fresh ideas and methods. I'm not too familiar with what you're doing, but as I was browsing my list of titles, I discovered that you are working in the same application domain as I am.

Solution seekers will read the method section, the theoretical section, and anything else that can help them. They could be surgeons looking for artery modeling software, or AIDS researchers who heard that small-world networks have interesting applications in their field. Their knowledge gap may be very large. They are looking for general articles or even specific articles, which they will read in part, expecting to find a clear but substantial introduction with many references to further their education.

The young researcher

Hello! I'm fresh out of university, and quite new to this field. Your paper looks like a review paper. That's exactly what I need right now. Nothing too complicated. Just enough for me to understand the field, its problems, and the solutions advocated by researchers. That will do just fine!

Young researchers will read the introduction and (perhaps) follow your trail of references. They do not expect to be able to make sense of everything the first time, but what little they can understand, they will be happy with. Their knowledge gap is great.

The serendipitous reader

Hi! Cute title you've got there. I had to read your paper. Such a title could only come from an interesting writer. I thought I would learn a few things, a paradigm shift maybe. I'm not sure that I will understand any of it, but it's worth a try. Last time I did that, I learned quite a lot. The paper had won the Best Paper Award in an IEEE competition. I studied the paper. Although I did not understand much, I got quite a few hints on how to improve my scientific writing skills!

My point is this: researchers will come to your paper with different motivations and needs. A common mistake is to imagine the reader as another you, the competitor in this story or someone who knows your topic as well as yourself. As the author, you would be wise not to rush through the introduction and the list of references. You would also be wise to provide enough detail so that other researchers can check and validate your work: 'little validation, little value'.

Ask people to read your paper. Ask for their opinion. Is it written for experts like yourself or will researchers new to the field be able to benefit from your paper? Are they motivated to read the rest of your paper after reading your introduction? Ask readers to circle the parts of your paper they found difficult to understand. Don't ask them to explain why, though. They probably would not give you a good answer. But then examine each of these circled parts and revise them drastically.

Chapter 8

Bridge the Knowledge Gap

> Apple computer
>
> *In the past, the computer Gods in the mainframe world could only be approached by the grand priests in white coats serving them. When the personal computer arrived, the Gods did not quite fall off their pedestals. They just moved from the computer room temple to the living room shrine. The computers had tamed their owners. Occasionally, through what looked like sorcery to humans, some PC owners managed to tame their computers. Application programs responded with docility to the secret incantations they typed. That knowledge, known only to them, had corrupted their virgin brains. They no longer remembered what it was like not to know. Then came the Macintosh, to give hope again to the rest of the human race. History tells us the Mac dispelled the Orwellian vision of 1984, but it did not quite manage to allay the fears of the uninitiated the way the iPod and the iPad did.*

Crossing the knowledge chasm between you and your reader is not easy. The reader knows less than you, but how much less?

- It depends on you. If the new knowledge you are contributing is significant, the knowledge gap between you and your readers is large.
- It depends on your readers. If their own knowledge of your field is small, they may not be familiar with the vocabulary or methods used. As a result, their initial knowledge gap is large, even if the additional knowledge you bring is modest.

You need to evaluate the gap to make sure your paper reaches the readers described in the previous chapter: the field intelligence gatherer, the competitor, the problem or solution seeker, and the young researcher. Of course, you could assume that the reader is knowledgeable enough to follow your paper, but is this assumption valid? What do you, the writer, **know for sure** about your readers' initial knowledge?

You know that your readers found interesting one or several keywords in your title. You know that your readers have *enough* knowledge to tackle and explore *parts* of your paper. You know that your readers read the journal in which your paper is published or attend the conference where your paper is presented. You know that their work is related to the domain covered by the journal or conference. For example, readers who attend the International Symposium on Industrial Crystallization are in chemical engineering. They know the tools and techniques used in the domain. They know the meaning of centrifugation, phase separation, concentration, calorimeter, and polarized light microscope. They know the principles of science, how to conduct experiments, and how to read a concentration and temperature chart. They know English — the Queen's English, the President's English, or some flavor of broken English.

Now that we have ascertained what you know for sure about your readers' initial knowledge, what then do you **not know for sure**?

The answer is short and simple: **everything else**. Indeed, everything else cannot be assumed to be known. Even though it is tempting to believe that readers have the same level of knowledge as the one you had at the start of your project, nothing could be further from the truth. Readers are not younger versions of you.

Now that we have looked at their initial knowledge, let us consider your contribution. How much do they know about it? Nothing! Your contribution is unknown to them, **just as it was unknown to you before you started the research that led to your paper.**

Let us suppose that the title of your article is the following:

"Phase transitions in lysozyme solutions characterized by differential scanning calorimetry."[1]

Some readers may be more familiar with characterization techniques than they are with lysozymes. Therefore, they do not know which

---

[1] Lu J, Chow PS, Carpenter K. (2003). Phase transitions in lysozyme solutions characterized by differential scanning calorimetry. *Progress in Crystal Growth and Characterization of Materials* **46**: 105–129.

data, method, or experiment best applies to lysozyme, nor would they know what others before you have done in this domain or what specific problems remain unsolved. That is precisely what they will discover while reading your article.

Bridge to Ground Zero

I hope you now see that, on the whole, the gap between your elevated knowledge and your readers' basic knowledge (ground zero) is wide.

Since it is impossible to guess how wide the gap is, you will have to set the minimum knowledge required, the reasonable ground zero on which you will build new knowledge. It would be unreasonable to write your paper for college students or for scientists who are not regular readers of the journal your paper targets.

To put it in an easy to remember formula:

Reader Knowledge Gap = The new knowledge you acquired during your project + The new basic knowledge the reader requires to reach your starting point.

Ground zero will be conditioned by the number of pages given to you by the journal. The more pages you have, the more you can lower ground zero or increase the size of your contribution. For short papers for example, you can settle for a higher ground zero or a smaller contribution, and you can use the short reference section of your paper as a knowledge bridge. References are convenient shortcuts that tell the non-expert reader *"I'm not going to explain hidden Markov models. Indeed, I'm going to use the acronym HMM when I refer to them. This should be common knowledge to anyone working in the domain of speech recognition. Go and read reference [6] that represents the classical work on the subject if you need more basic information."*

Assuming the editor gives you enough pages, you may decide to lower ground zero by providing extra background knowledge instead of asking readers to get up to speed by themselves. Writers often provide this knowledge in **a background knowledge section** that immediately follows the introduction. This section is a great place to summarize what readers would have learned had they had the time to read the articles you mention in your reference section (you can safely assume they will not read your [1], [2], and [3] before reading the rest of your paper). This background knowledge section is not part of your contribution, but it is necessary to understand it.

Ground zero is set by the latest books or review articles written by domain experts. If such articles are not available, then a look at the latest conference proceedings in your domain area should give you a couple of general articles that will accomplish the same function. Ground zero keeps moving up. Science is built on science, and scientists are expected to keep abreast of what is happening in their domain.

Ground zero is set by the type of journal. Some journals are multidisciplinary. The journal of bioinformatics, for example, is read by two distinct types or readers with widely different backgrounds: computer scientists and biologists. Some journals have such a broad audience (*Nature* for example) that the whole paper has to be written for non-experts. Just-in-time background is extensive in these journals.

Bridge to Title Words

Readers have chosen to read your paper because certain keywords in its title attracted their attention. This very fact is of utmost importance. Because of it, and as incredible as it sounds, the knowledge gap of the reader is predictable. Each specific keyword in your title attracts two types of readers: the readers who are very familiar with the keyword and do not require any background, and the readers who would like to know more about that keyword and who indeed require background. Identify these keywords, and you have an idea of the background you need to cover in your paper. It is that simple.

Let us use a sample title to identify its potential readers. But before we start, some background might be helpful. To manufacture a non-stick frying pan, you need to deposit a hard coating on top of the metal pan. That coating is hydrophobic, which simply means water-repellent. The process to deposit the coating is called the sol-gel process. The *sol* is a chemical solution serving as a precursor which, after it enters into reaction with the *gel* made of polymers, will form the hydrophobic hard coating.

"Hydrophobic property of sol-gel hard coatings*"

What are the keywords?

1) Sol-gel: a subset of technologies for making coatings
2) Hard coatings: a subset of surface coatings
3) Hydrophobic property: a subset of the properties of surface coatings such as hardness, etc...

The knowledge field presented in this paper is established by these three keywords. Only experts among the readers have enough knowledge in all three. It should come as no surprise to find among the readers people with knowledge lacking in one or more of these keywords. And it should also be expected that, given the name of the conference (Technological Advances of Thin Films and Surface Coatings), all participants are somewhat familiar with the main knowledge sets: coatings, coating technologies, or coating properties.

Now ask yourself what these readers would like to know, and which part of your paper is going to provide them with an answer.

Hydrophobic property: What in the sol-gel process influences the hydrophobicity of hard coating? How is hydrophobicity related to other properties of the coating such as surface structure, transparency, or mechanical properties? How is hydrophobicity measured and quantified?

Hard coatings: How are hard coatings made by sol-gel technology? How does hydrophobicity affect or confer the hardness of the coating? How hard is the coating made of sol-gel material?

Sol-Gel: What are the sol-gel precursor materials used? How are the sol-gel coatings solutions prepared? What are the coating processes used?

Here is what should NOT be covered in the background because it falls outside the scope of your paper: hydrophobicity of soft coatings, other properties of materials independent of hydrophobicity, or hard coatings produced by other processes than sol-gel.

---

* The paper was presented by its author, Dr. Linda Wu, at the 2nd International Conference on Technological Advances of Thin Films and Surface Coatings (Thin Films 2004) in Singapore.

In this exercise, the title allowed you to identify five overlapping reader profiles, with different background needs:

1) People interested in sol-gel
2) People interested in hard coatings but only vaguely familiar with sol-gel
3) People interested in the hydrophobic properties of any coating material hard or soft
4) People interested in the mechanical properties of sol-gel hard coatings eager to know if the mechanical properties are compromised by the hydrophobic properties of the coating
5) People interested in the hydrophobic properties of sol-gel hard coatings (your contribution).

Now that you understand the principle, conduct the same exercise on the title of your own paper.

✳ Before you write your introduction, use the title keywords to identify who your readers are, and provide the background they probably need to understand and benefit from your paper.

Just-in-Time Bridge by Way of Local Background

In the chapter on memory, we determined that there were two types of background: the global background found in the introductory parts of your paper, and the local background found wherever just-in-time information is needed. The local background includes the definition of highly specific keywords that only experts use. Problem is, you no longer realize that non-experts don't know these keywords. To use a metaphor, imagine yourself standing next to your reader at the foot of a hot-air balloon. You are about to embark on your research. You climb inside the gondola. As research progresses, the landscape you see is different from the one seen by the reader on the ground. Your discoveries raise your knowledge level as you drop the sandbags of your ignorance. That ignorance now becomes your reader's ignorance. Your balloon rises slowly (or rapidly, for one does not usually dictate the pace of discoveries). By the time you are ready to write, you have risen far above the reader, as in ☛1.

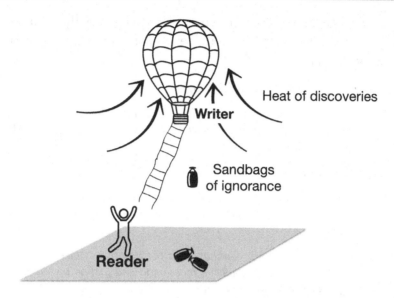

Figure ☞ 1
Knowledge elevates the writer above the reader.

Your job as a writer is to bridge the gap created through months of research. You have to throw a figurative ladder to readers so that they can come on board your hot air balloon gondola. How far down should you throw the ladder? Down to ground zero. Often the ladder is too short. Ground zero is not properly identified — the background is for tall experts only. Or, it may be that ground zero is properly identified, but rungs on the ladder are missing. These rungs correspond to the local background. Readers remain suspended in mid-air, stuck in the middle of a section of your paper, frustrated, trying to get on board but unable to climb further — you skipped an essential logical step or piece of background information that prevents them from completely benefiting from your paper, or you used a new word they do not understand.

Asking a reader to be familiar with reference [X] before being able to understand the rest of your paper is equivalent to asking the reader to get off the ladder, go to the library, read *the whole article* [X], climb the ladder again, add the missing rung, and keep on climbing.

☀ It would be more advisable, space allowing, to briefly summarize in your paper whatever reference [X] contains that is of interest to the reader.

The authors could have written this:

> *The dynamic behavior was expressed in the Unified Modeling Language (UML, Booch et al. (1999)). The notation used in Figure 3 is that of UML sequence diagrams.*

Instead, they wrote this (note the footnote "*")

> *The dynamic behavior was expressed in the Unified Modeling Language (UML, Booch et al. (1999)). The notation* used in Figure 3 is that of UML sequence diagrams.*

> *(in footnote) For those not familiar with the notation: Objects line up the top of the diagram. An object's messaging and lifeline boundary is shown by a vertical dashed line starting below the object. Object activity is shown by the activation bar, a vertical rectangle drawn along the lifeline. Horizontal arrows issued from a sender object and pointed to a receiver object represent the messages sent.*

There is no need to build an elaborate rung on your ladder. Your goal is not to have the reader marvel at the exquisite design of a ladder rung, its elaborate anti-skidding grooves, its ornate structure... The role of the ladder is to take your reader inside the gondola of the air balloon where your contribution is revealed and understood. So here are the parting words at the end of this chapter. They come from biologist and writer, Thomas Henry Huxley.

"The rung of a ladder was never meant to rest upon, but only to hold a man's foot long enough to enable him to put the other somewhat higher."

 Read your paper. Is your introduction too short? Is it motivating? Have you identified a ground zero that is reasonable to expect from your reader? Have you identified the background the keywords of your title make necessary? Have you identified the intermediate discoveries that removed the sandbags of your ignorance and elevated your knowledge above that of the reader? Should these constitute part of your background?

Chapter 9

Set the Reader's Expectations

> **The train**
>
> *Imagine the mind of the reader as a train. The author provides the set of tracks and the signal boxes. What could go wrong?*
>
> 1. *No tracks—Readers, left on their own, have to pick the loose rails lying around to create their own set of tracks. Very slow going.*
> 2. *No signal box—expectations are not set, so readers chug along slowly, not knowing where to go next.*
> 3. *Faulty signal—Readers, misled, are going down the wrong tracks.*
> 4. *The train is in a tunnel—Readers tolerate being left in the dark for a short while, as long as the end of the tunnel is in sight, and clarity returns.*

Creating and fulfilling the reader's expectations guarantee that reading will be fluid, interesting, and fast. That is why setting expectations is such an essential technique. In this chapter we will discover what sets expectations, and once we know, we will review how to set them. But first, we need to refresh some grammatical knowledge because grammar plays a role in the alignment of thoughts between reader and writer. A well-written sentence sets clear expectations that align the reader's thoughts and feelings with the writer's thoughts and feelings. Alignment is necessary to avoid distortions, misinterpretations, or ambiguity.

Expectations from Grammar

Main clause–subordinate clause

A **clause** contains a subject and a verb.

Learning needs to be semi-supervised.

Subject   Verb

A **subordinate clause** does not stand alone; to be understood, it needs a **Main Clause**.

Subordinate clause → *Because variation within each class is large,* Main clause → *learning needs to be semi-supervised.*

A sentence may have two **independent clauses**, which stand alone. They are separated by punctuation and often have a contrasting hinge (*'however'* and *'but'*).

Independent clause Generally, supervision is required Punctuation; Other independent clause however, for classes with low variations it is not required.

Independent clause Generally, supervision is required Punctuation, Other independent clause but it is not required for classes with low variations.

These independent clauses are kept within the same sentence to establish the contrast between required and not required. Note that they share the same subject (*'supervision'*).

The placement of words in a sentence influences expectations

Do you want to find out how grammar sets your expectations? I suggest that we take a small survey together. Use a pencil if you wish to be able to erase the marks after the survey. You will see four sentences, each one on the topic of *Evolutionary Algorithms* — computer techniques based on biological evolution to find the optimal solution to some problems. After slowly reading each sentence, mark down how you feel about them. Do not re-read multiple times or ponder the matter at length before answering, because readers generate impressions "on the fly," as they read. One last piece of advice: if you want to make sure that the answer to a question does not influence the answer to the next question, take a short pause, look away from the book, take a sip of water, and return to the next sentence.

Feel good about Evolutionary Algorithms? Tick the box under the happy face ☺

Feel neutral about Evolutionary Algorithms? Tick the box under the neutral face ☺

Feel negative about Evolutionary Algorithms? Tick the box under the sad face ☹

(1) Evolutionary Algorithms are sufficiently complex to act as robust and adaptive search techniques; however, they are simplistic from a biologist's point of view.

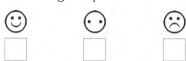

(2) "Evolutionary Algorithms are simplistic from a biologist's point of view, but they are sufficiently complex to act as robust and adaptive search techniques."[1]

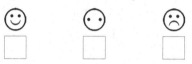

(3) Although Evolutionary Algorithms are sufficiently complex to act as robust and adaptive search techniques, they are simplistic from a biologist's point of view.

(4) Evolutionary Algorithms are simplistic from a biologist's point of view, although they are sufficiently complex to act as robust and adaptive search techniques.

Now look back to your answers. Did you tick the same box for each sentence? If not, why not?

All four sentences present two identical facts: (1) Evolutionary Algorithms are simplistic from one point of view, and (2) Evolutionary Algorithms are sufficiently complex to act as robust and adaptive

---

[1] *(2) is reprinted with permission from Sinclair M. (2001) Evolutionary Algorithms for optical network design: a genetic-algorithm / heuristic hybrid approach. (name of university). Thesis.*

search techniques from another point of view. If you agree with one point of view (for example, if you find them simplistic), then all sentences 1 to 4 should carry the same message and be equally perceived, whatever the order of the words in the sentence. Yet, this is not the case, is it! You find some sentences favorable and others unfavorable. Something influences the way you perceive the facts presented. What is it? The answer is found in the grammar.

From a grammatical perspective, sentences 1 and 2 are similar in that they both have two independent clauses; what changes is the order of the clauses.

Looking at the opinions of a panel of 33 readers in Table ☜1, one sees the strong polarization of answers between the first two sentences: Sentence 1 is predominantly neutral to negative whereas sentence 2 is predominantly positive.

Table ☜1
Influence of the placement of the information in a sentence

	Sent.start	Sent.end	Main	Sub.	Neg.	Neutral	Pos.
Sentence 1	Pos.	**Neg.**	Pos/Neg		**15**	14	4
Sentence 2	Neg.	**Pos.**	Neg/Pos		3	4	**26**
Sentence 3	Pos.	**Neg.**	Neg.	Pos.	**22**	9	2
Sentence 4	Neg.	**Pos.**	**Neg.**	Pos.	12	**19**	2

To understand what happens, let's sort the sentences according to two criteria: the majority vote of the panel, and the way the Evolutionary Algorithms are featured at the end of each sentence — simplistic being negative and sufficiently complex being positive

Sentence (1): Panel ☹ and end sentence ☹   ✓ MATCH
Sentence (2): Panel ☺ and end sentence ☺   ✓ MATCH
Sentence (3): Panel ☹ and end sentence ☹   ✓ MATCH

People who took the survey seem to align their opinion on the opinion expressed at the end of each sentence.

☀ To encourage the reader to agree with you on what is important, place the information you consider important at the end of a sentence.

We have not yet looked at sentence 4.

Sentence (4): Panel ☺ and end sentence ☹   ✘ NO MATCH

Why doesn't sentence (4) fully behave like the other three? What changes in the sentence to change the views of the panel?

Evolutionary Algorithms are simplistic from a biologist's point of view ← **Main clause, Subordinate clause** → *although they are sufficiently complex to act as robust and adaptive search techniques.*

The sentence the panel sees as neutral, represents a balance of opinions. It seems to be the result of a fight between two sources of influence with a bias towards the negative opinion. The negative point of view is expressed in the main clause and it seems that the panel of readers is influenced more by what comes in the main clause than by what comes in the subordinate clause. The reason people are indecisive is explained by the fight for influence:

Evolutionary Algorithms are simplistic from a biologist's point of view is in a weak position at the beginning of the sentence, but reinforced by the strong influence of the main clause →; and *although they are sufficiently complex to act as robust* and *adaptive search techniques* has a strong position at the end of sentence (+) tempered by the weak influence of the subordinated clause (−). They balance each other.

If what we suggest is valid, then sentence (3), which like sentence (4) has a main and subordinate clause, should verify the strong role of the main clause, and it does. In sentence (3), the information that gets the most votes from the panel is in the main clause.

☀ To encourage the reader to agree with you on what is important, place the information you consider important in the main clause.

But why stop here! We can extract more interesting guidelines from this fruitful example. You may have noticed in Table ☛1 that the number of indecisive people is larger for sentence (1) than for sentence (2). Why is that?

It may be a combination of two factors: punctuation and the diluted strength of the adverb '*however*'. Whenever the brain encounters a punctuation like the period or the semi-colon, it stops and processes the meaning of the sentence. In the case of sentence (1), the brain gets a favorable impression of evolutionary algorithms at the semicolon pause. The second part of the sentence, starting with 'however' in the strong position at the end of sentence, attempts to reverse that already very positive impression. It succeeds partly (negative score greater than positive score), but not totally. The reason for that partial failure is due, in my opinion, to "however', an adverb overused to bring contrast, and often used wrongly to change topic (fake contrast). As a result the people who were positive only move to the neutral attitude, whereas the people who were neutral remain neutral or turn negative.

For sentence (2), the situation is quite different. It is very positive. Why? Again, it may be a combination of two factors: punctuation and the undiluted strength of the conjunction '*but*'. There is no semantic closure after a plain comma, the brain continues reading before coming to that final closure at the end of the sentence where the strong positive impression is formed. To reinforce that positive impression, the hammer-like '*but*' rebuttal squashes the negative impressions.

☀ To encourage the reader to agree with you on what is important, place the information you consider important after the conjunction 'but', and/or right before a period or a semi-colon.

There is still the matter of variability in the answers of the panel — the people who did not go with the majority vote. What influenced their opinions? There are several factors, which I list here.

1) People's temperament and outlook on life. Some are influenced more by what is positive than by what is negative.

2) People's jobs. A biologist may place emphasis on the negative statement whereas a computer scientist may have more positive feelings for the algorithm.

3) Some people consistently voted against the majority vote. I discovered that those who voted this way were non native English speakers whose native language was based on Sanskrit. In their grammar, the important information is mostly placed at the beginning of a sentence, not at the end. So be aware that if you are in this situation after taking this test, your foreign grammar influences the way you write English, and that may confuse readers.

4) Your answer to the first question may bias your answer to the second question, and so on (remanence effect). That is why I regularly change the sentence order when conducting the test. The good news is (as far as you are concerned) that the guidelines stated here are confirmed by the 280 scientists who took the test.

☀ Consider starting sentences with a subordinate clause so that they end with a convincing main clause.

Forget about Mrs. Smith, your English teacher who told you to "**Never** start a sentence with **BeCause**", and she stressed every syllable to show you she meant it. Of course, had you asked "Why, Mrs. Smith?" She would probably have responded: "Because I say so".

The following words create a subordinate clause at the head of a sentence so that the main clause can end the sentence. They also make excellent attention-getters, and really shine at setting expectations.

Because...If...Since...Given that...When...Although...Instead of...While...

Take a word like *because*. Placed at the beginning of a sentence, *because* announces an upcoming consequence in the main clause. It sets a delay between the time the expectation is raised and the time it is fulfilled. That delay creates tension and momentum. The tension acts like a metallic spring: it pulls reading forward. In the real world, the length of a spring matters less than its strength; likewise, a sentence's length matters less than the tension created by the arrangement of its words.

☀ To build suspense within a sentence, start the sentence with a subordinate clause. This creates a tension that the end of the sentence releases.

Dynamic sentences have words that set expectations

Which words create expectations in the next sentence?

Up to this point, we have only considered basic filtration techniques.

If you answered: the locution *up to this point*, the adverb *only,* and the adjective *basic,* you were absolutely right. *Only* shows that more is coming. *Up to this point* works in tandem with *basic*. Together, they confirm that basic filtering techniques are no longer going to be covered. More advanced filtering techniques will now be considered. How do we know that? Take away the adjective *basic*.

Up to this point, we have only considered filtration techniques.

With *basic* missing, the reader expects that we are leaving the topic of filtration techniques to look at other techniques. So it is the adjective *basic* that sets expectations. Now test yourself with the next sentence. Which are the words that generate expectations?

Dengue fever epidemic in India does not occur at the beginning of the monsoon season.

The words that set expectations are the negation *does not* and the noun *beginning*.

1) Expectation of explanation — Why doesn't it occur at the beginning?
2) Expectation of elaboration — When exactly does it start?

This sentence indirectly tells us that dengue fever epidemic occurs, not at the beginning but sometimes *during* the Monsoon season. Its purpose is to emphasize when it starts. The writer prepares the reader to an explanation of the conditions for the epidemic to take place.

☀ Adjectives, adverbs, and nouns associated with a negation or with a pejorative meaning, allow fast transfer of expectations to their opposites.

Compare contrasted (A) with flat (B).

*(A) Trapping is unimportant at high temperatures where there is plenty of energy to escape. But **trapping** leads to very slow dynamics at low temperature.*

*(B) **Trapping** is important at low temperature because it leads to very slow dynamics, as there is not much energy for the molecules to escape.*

1) The total number of words for both texts is identical: 24 words. Paragraph (A) has two sentences: the first has one main clause and one subordinate clause, and the second has one independent clause. Each sentence is short (14 and 10 words). Sentence (B) is long (24 words) and the main clause has two cascaded subordinated clauses ("because", "as") making it more complex.

☀ Sentences that end with nested sub-clauses create weak expectations and they move further and further away from the main topic of the sentence.

2) With (A), the reader is alerted on the role of low temperatures. Sentence (B) is less contrasted because low and high temperatures are no longer compared; it creates the expectation that the next sentence will keep the same topic, i.e. trapping. Sentence (B) does not create the same compelling

wish to know more about "very slow dynamics" because it is tucked away in the middle of the long sentence.

☀ Unguided, the default expectation is that the topic does not change from one sentence to the next.

3) But the punctuation has changed. The full stop in the original (A) has been replaced by a comma in (B). The full stop is better. It gives more breathing space than the comma. During this very useful breathing space, the brain processes the sentence and lets its meaning sink in. Based on this fresh understanding, the brain then sets expectations for the next sentence.

☀ Place the punctuation where it creates expectations.

Cognitive neuroimaging

Michael works in a cognitive neuroscience laboratory. He explores the brain with functional MRI, and seeks to understand what happens in our working memory. I ask him what happens when we read. Michael, an extremely well organized man, retrieves from his computer two papers from Peter Hagoort: "Integration of Word Meaning and World Knowledge in Language Comprehension," and "How the brain solves the binding problem for language: a neuro-computational model of syntactic processing."

Somewhat intimidated by the titles, I ask if he would not mind explaining in layperson's terms what happens when we read. Still facing his Macbook Pro, he quickly thinks and asks, "Do you use Spotlight?" I reply, "Of course." Any Mac owner is familiar with the search function of Spotlight, the little white magnifying glass located in the top right corner of the Mac menu bar. "Look here," he says. I get closer to his screen. "As I type each letter in Hagoort's name, the search engine immediately updates the search results. H, then HA, then HAG. Notice how the list is now very small; one more letter and we will have zoomed down to Hagoort's papers."

As soon as he types the letter O, the list shrinks down to a few items, and among them, Hagoort's papers. He turns towards me as I sit back into the chair facing his desk. "You see," he says, "it looks as though the Mac tries to guess what you are looking for. Similarly, while you read, your brain is active, forever seeking where the author is going with his sentence. It analyses both syntax and meaning at the same time, going from one to the other transparently."

Expectations from Science

Verbs, adjectives and adverbs make claims

The next example combines expectations from science and expectations from grammar.

> *Tom Smith's assumption [4] that no top layer material could come from the by-products of the pinhole corrosion which had migrated is not supported by our data.*

Despite all its problems, that sentence still manages to create an expectation. A scientific claim is made: the data **does not support** Tom Smith's assumption. As scientists, we expect the writer to support his claim with data.

Here is the sentence again, but this time with verb closer to subject, and with added dynamism thanks to the active voice. This sentence is written in the present tense, a tense usually used for claims. The auxiliary "*do*" shows certainty.

> *Our data reveal that, contrary to Tom Smith's assumption [4], the pinhole corrosion by-products do migrate to form part of the top layer material.*

Have the expectations changed? What do you expect the writer to introduce in the next sentence? The data or Tom Smith? The majority of you will answer, "The data". There is a slight difference with the first sentence, though: the expectations are more defined now. The reader expects proof of the claim regarding migration, or proof that the material found on the top layer comes from the pinhole. Tom Smith is now between commas, a side remark, and nobody cares about his assumption. The findings of the author are stated, not those of Tom Smith.

But imagine for an instant that you are the writer and you want to focus the reader's attention, not on the data, but on Tom Smith's assumption. How would you rewrite the previous sentence?

There are two ways of rewriting the sentence; both have in common the placement of Tom Smith at the end of the sentence.

> *Our data reveal that the pinhole corrosion by-products migrate to become part of the top layer material, contrary to Smith's assumption [4].*

Our data reveal that the pinhole corrosion by-products migrate to become part of the top layer material. These findings contradict Tom Smith's assumption [4].

In this last sentence, Tom Smith's assumption is no longer a side remark, it is the main point and it comes in a small package: a short punchy sentence. The reader is now curious. What did Tom Smith assume? Why is there a contradiction? We know what the findings of the writer are. What we do not know clearly is what Tom Smith assumed.

☀ Expectations of change are mostly set by the new information found at the very end of a sentence.

So far, we have looked at claims made by verbs (*'is not supported,' 'do migrate'*). Adjectives and adverbs also make claims. Readers of scientific papers have different expectations than readers of novels. When a novelist writes: "the ferocious dog", the reader's imagination recreates the image of this dog from past face-to-face encounters or from movies where such dogs are seen. It may not be the type of ferocious dog the writer had in mind, but who cares — the more ferocious the reader makes it, the better! Unlike the scientist, the novelist does not have to convince the reader of the dog's ferocity by measuring the surface of the barred dental area, the number of milliliters of saliva secreted per minute, or the dog's pupil dilation — all this on live ferocious dogs of course. *In vivo* data collection is a dangerous occupation, especially if you want your population to be representative, and go beyond the Chihuahua to include German Shepherds, Doberman Pinschers, and Rottweilers.

Adjectives or adverbs are subjective. What is robust to you may be fragile to me. What is very fast to you may be moderately fast to me. In Science, adjectives and adverbs are claims. When the claim stated is widely accepted as true, the scientist does not expect the writer to justify it. But remember, what is obvious to an expert, may not be obvious to a non-expert who might still want you to justify your adjectival claim. It is safer to justify than to assume no justification is needed. And it is safer to define an adjective than to let the reader define it.

When feed gas changes from light to heavy gas, the plant load decreases by a small fraction (4.7%).

In this example, the adjective *small* is immediately quantified. Elsewhere in the paper is a table that gives the definition of what constitutes heavy, medium, and light feed gas, using feed gas components (mole fraction) expressed in % or ppm.

☀ Quantify adjectives when possible or define them.

 Read your abstract and your introduction. Highlight all adjectives in fluorescent yellow, and adverbs in fluorescent red. If your paper glows in the dark, then you have work to do. Examine each adjective and adverb. Are the claims justified? Would removing an adjective make you more authoritative? Could each adjective be replaced by a fact?

The main sections of a paper create their own expectations

The next few examples show that expectations are also guided by the part of the paper the sentences appear in.

This sentence, the first in the introduction, makes an adverbial and adjectival statement that creates a strong expectation regarding the goal of the paper.

'Traditionally, airplane engine maintenance has been <u>labor-intensive</u>.'

Both "Traditionally" and "has been" confirm that the labor intensive situation is as true today as it was in the past. Expectations for the paper seem pretty clear: the topic of the paper is about airplane engine maintenance, and the reader assumes that the author has found a novel way to make airplane engine maintenance less labor-intensive by using robots, by reducing the number of parts, or by improving the quality and durability of airplane engine components. But wait! Why should the reader assume that the paper's contribution is to reduce labor? It may be because of the word "traditionally". In the context of a scientific paper, the 'traditional' way of doing things is akin to saying the "old" or "low-tech" way. But in the context of a ethnographic paper, "traditionally" could instead imply a custom worth preserving! The point here is, some expectations come not from the grammar and vocabulary, but directly from the existing contextual knowledge of the reader. In the context of a modernized country, labour-intensive = high-costs = bad. Whereas in the context of a poor country with high unemployment, labour-intensive = many employment opportunities = good.

☀ Context of statement and general culture shape expectations.

This other example from the methodology section of a scientific paper establishes expectations around a time-ordered sequence. The clue that a time sequence is in progress is clear. Each sentence starts with, or contains, an expression related to time: 'day one,' 'next three days,' 'after,' 'over the next...'.

> On day one, thirty embryonic cells are placed in the culture dish. For the next three days they are left to proliferate. After proliferation, the cells are collected and put into new culture dishes, a process called replating. Over the next three weeks, 180 such replatings produce millions of normal and still undifferentiated embryonic cells.

☀ Expectations of sequential progression require sequence markers such as time markers, verb tense, or consecutive numbers. In this case, expectations are set by the first words of the sentence (not its last words).

In this final example, an abridged abstract, the author laid out the sentences according to a specific order expected by the reader: that of a scientific paper: aim, methodology, result, and significance, which also corresponds to that of the scientific process: observation, hypothesis, experiment, result and interpretation.

> **[Observation]** The dengue genome forms a circle prior to replication, as is the case for the rotavirus. Since one end of the circling loop is at the 3′ end of the genome where replication takes place, **[Hypothesis]** we wondered if the loop had an active role to play in the replication. **[Experiment]** After comparing the RNA synthesis capability of various whole and truncated dengue genomes using radio-labelled replication arrays, **[Results]** we found that a region other than the 3′ end of the genome had an even larger role to play in the replication: the 5′ end of the genome. Although far away from the 3′ end, it seems to loop back into it. **[Discussion]** Thus, it may be that the promoter site for

RNA synthesis resides in this unusual location. Looping would then be a means of bringing the promoter to where it can catalyze rapid duplication.

☀ Expectations of logical progression require sequence markers such as logical markers ('if,' 'then,' 'thus,' 'therefore'), verb tense or auxiliary to express suggestions ('would,' 'could,' 'may'), or consecutive numbers expressing logical steps which are not necessarily based on a time sequence.

Chapter 10

Set Progression Tracks for Fluid Reading

When readers cruise down your paper in fifth gear, it is because you have created a highway for their thoughts to travel on at great speed, a highway that stops their mind from wandering where it should not go. Sometimes, while reading certain papers, I feel as if I am driving in the fog at a crawling speed across a muddy field, trying to follow somebody else's tracks. In Science, unlike literature, you guide your readers along a clearly lit well-signposted highway. How to build such a highway is described in one word: progression.

Progression is the process of transforming what is new into what is known. It builds a coherent context that allows readers to travel light and read on with minimum cognitive baggage. When readers start a sentence, a paragraph, or a section in your paper, they relate what they read to what they know. This progressive anchoring of new knowledge onto old knowledge is an essential learning mechanism.

The reader may wonder if there is a connection between expectation and progression. Sometimes they are joined and work in synergy. For example, when a sentence starts with *first*, it sets the expectation that the following sentences will cover what comes after; the sequential progression fulfills the expectation. Sometimes expectation and progression are distinct and work separately, hopefully in the same direction. It does happen that they pull the reader in different directions, which is not advisable, of course. Progression should always support the expectation, not the opposite.

There are other differences between the two. Expectations have a longer reach. For example, a question's reach extends far beyond the next sentence. Progression is purely local: between two phrases within a sentence, two sentences within a paragraph, or two paragraphs within a section. Expectations open the reader's mind. Progression keeps the reader's mind on your rails.

But before we dive into progression schemes, we must first revisit some grammar. In France, schoolchildren discover all about topic and stress (in French *Thème* et *Propos*) in their grammar book at the age of 14. By the time they go to university, they have forgotten all about them! The same probably applies to you, so here is a quick refresher course.

Topic and Stress

"Mr. Johnson hurriedly chased his neighbour's dog out the front door".

This sentence contains three elements: a man named Johnson, a dog, and a door. If I were to ask you who or what this sentence was about, you would instinctively (and correctly answer) "the man, Mr. Johnson". Indeed it's his story — he is the topic of the sentence, positioned upfront, and the doer of the verb "chased".

Imagine I instead change the sentence to:

"The neighbour's dog sneaked into Mr. Johnson's house, but it wasn't long before the man hurriedly chased him out through the front door."

The topic of the sentence is now the dog — it is now the dog's story, and Johnson is a secondary actor.

Finally, compare both of the previous sentences to this one:

"The front door to Mr. Johnson's house burst open, the neighbour's dog running out quickly, Mr. Johnson right on its tail."

Despite containing roughly the same information, all three sentences are read from different perspectives, because they have different topics. The topic should always be placed upfront in a sentence. Anything that is not the topic is called the stress. Typically, the stress starts at the first verb and finishes the sentence.

| **Topic** > The neighbour's dog
| **Stress** > sneaked into Mr. Johnson's house [...] front door.

The topic establishes the main subject of the sentence, and the stress details what happens to that subject.

☀ To identify a topic, evaluate whether what it describes is known and is at the beginning of a sentence. The stress is new to the reader.

Inverted Topic

We've seen that the topic should come upfront in a sentence. What happens when it doesn't? The topic of the next two sentences is "the cropping", but it is placed wrongly in the second sentence.

The cropping process *should preserve all critical points. Images of the same size should also be produced* **by the cropping.**

These sentences do not seem well balanced. This is because, in the second sentence, the already known information (*the cropping*) is at the end in a stress position, a place traditionally reserved for new information, whereas the new information is in the place reserved for context. This inversion delays the understanding of the sentence until its topic finally arrives to clarify everything.

Here are three ways to correct the problem:

Change the voice in the sentence from passive to active or vice versa, thus straightening the inverted topic and stress by bringing the known information to the head of the sentence.

> *The cropping process should preserve all critical points. It should also produce images of the same size.*

Invert the order of the sentences to re-establish progression.

> *Images of the same size should be produced by the cropping. The cropping should also preserve all critical points.*

Combine the two sentences into one.

> *The cropping process should preserve all critical points and maintain the size of images.*

☀ Do not invert topic and stress in a sentence.

When the verb that starts the stress is strong, it may overshadow the rest of the stress and override weaker expectations.

> *Applying Kalman filters **reduced** the noise in the data sent by the low-cost ultrasonic motion sensors. **The reduction** was sufficient to bring down the detection error rate below 15%.*

☀ Use strong verbs to control expectations.

Topic Sentences

As we've seen, topics come upfront in a sentence and often introduce the subject of that sentence. What if you wanted to introduce the topic of not just one sentence, but one whole paragraph?

> *Black opals are extremely rare gemstones, making them highly sought after (and highly priced) in gem markets.*

What do you think will be covered in the subsequent paragraph? It could go into the rarity of black opals, their value in the gem market, or even just more general exciting facts about black opals. The sentence creates a very strong expectation that we will more deeply explore the topic of the sentence at length. This type of sentence is called a topic sentence.

Topic sentences always appear at the beginning of a new paragraph, and are fantastic ways to guide the reader's expectations over a longer period of time.

Three Topic-Based Progression Schemes to Make Reading Fluid

Progression around a constant topic

The scheme is straightforward. The subject of the sentence is repeated in successive sentences, directly or by a pronoun, or a more generic name. The reader is already familiar with the topic and reading is fluid. In this example, '*trapping*' (of the molecule) is the constant topic.

> "**Trapping** is unimportant at **high temperatures** where there is plenty of energy to escape. But **trapping** leads to very slow dynamics at **low temperature**. In the case of liquids, this **trapping** causes the glass transition — a dramatic slowing of motion on cooling."[1]

With the topic repeated often near the beginning of each sentence, reading is very fluid.

Topic to sub-topic progression

In this progression, the main topic is usually announced in the first sentence, and subsequent sentences examine aspects of the topic. In the following example, the first sentence is about visuals. The next two sentences review two aspects of visuals: their placement, and their convincing power.

> **Visuals** are star witnesses standing in the witness-box to convince a jury of readers of the worth of your contribution. **Their placement** in your paper is as critical as the timing lawyers choose to bring in their key witness. But most of all, **their convincing power** is far beyond that of text exhibits.

---

[1] Reprinted excerpt with permission from Wolynes PG. (2001) Landscapes, Funnels, Glasses, and Folding: From Metaphor to Software, *Proceedings of the American Philosophical Society* **145**: 555–563
[2] *ibid.*

Chain progression

In a chain progression, topic and stress are daisy-chained. The stress at the end of a sentence becomes the topic at the beginning of the next sentence. This frequently used progression scheme is easy for readers to follow. It is illustrated in the next paragraph.

> "The Brownian motion of the protein strand will carry it willy-nilly between various shapes, somehow finally getting it to settle down into a much less diverse family of shapes, which we call the **"native structure" of the protein**. The average **native structures of many proteins** have been inferred experimentally using X-ray crystallography or NMR."[2]

The elements in a daisy chain do not need to be repeated word for word from one sentence to the next. Often, the verb in a previous sentence becomes a noun at the head of the next sentence.

> Applying Kalman filters **reduced** the noise in the data sent by the low-cost ultrasonic motion sensors. **The reduction** was sufficient to bring down the detection error rate below 15%.

Sometimes, part of the previous sentence (bolded in the following paragraph) is brought forward in the next sentence by way of a pronoun such as 'This'.

> "The above observations can be generalized to a rather important conclusion. If **large mole differences between species exist in a data set** (and **this** is often the normal case for catalytic reactions), then the reactions involving both major and minor species should be rewritten to include only the latter. **This** should solve the problem of abnormal gradients in the extent of reactions for most cases."[3]

Don't fall in love with this paragraph though. It has its problems. It could be more concise: the *rather* in 'rather important' is unnecessary. It also contains a wonderful example of an ambiguous pro-

---

[3] Widjaja E, Li C, Garland M. (2004) Algebraic system identification for a homogeneous catalyzed reaction: application to the rhodium-catalyzed hydroformylation of alkenes using *in situ* FTIR spectroscopy. *Journal of Catalysis* **223**: 278–289, with permission from Elsevier.

Topic-based progression schemes

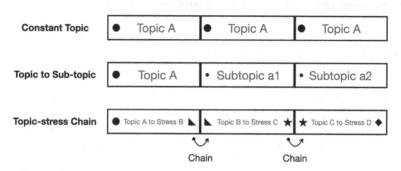

Figure ◀ 1
Three topic-based progression schemes. A sentence is represented by a rectangle; its topic by an object at the head of the sentence. In the chain progression, the topic retains the same shape as the previous stress to indicate that the stress of one sentence becomes the topic of the next sentence.

noun in the last sentence. What does '*This*' represent? If you answer '*the reactions*,' read the sentence again because it is not the correct answer. '*This*' refers to the rewriting.

Each of the three topic-progression schemes (Fig. 1) influences expectations and vice versa.

The constant topic progression answers the need to know more about the same topic (expectation of elaboration — breadth). The topic/subtopic progression answers the need to go deeper into the topic (expectation of elaboration — depth). The linear progression answers the need to see how things are related (expectations of relatedness and outcome). Going back to our metaphor of text as train tracks, the constant topic is the simplest scenario: the train continues in a straight line with no deviations. The topic-to-subtopic scenario takes the main train of thought on a sidetrack parallel to the main track. And the chain progression takes the train away from the main track to new destinations.

Topic-based progression schemes are complemented by schemes that do not depend on the topic.

Non Topic-Based Progression Schemes

Progression through explanation and illustration

The second sentence in the following paragraph introduces a progression of a new type: the **explanation**.

It usually follows a question or a statement that acts as a question.

Although the new algorithms have worst-case polynomial complexity, in practice they are shown to perform faster than the linear time algorithm. This is because their construction methods have good locality of reference and they have taken the computer's memory architecture into consideration.[4]

In this example, the first sentence says: the new algorithms should have performed worse than the existing ones, but they outperform them. The reader is intrigued and waiting for an explanation. It comes in the second sentence.

You can also announce the progression with transition words such as *for example, similarly, similar to, an analogy for, etc.,* which prepare an explanation based on a related example, an analogy, or a metaphor. Finally, you can illustrate with visuals. Just write (Figure #)!

What! Metaphors?

These last two chapters seem to discourage calling on the reader's imagination. One must define ferocity. One must keep the reader on track and channel reader thoughts. Does this mean that, you, the writer scientist, should keep reader imagination at bay because Science is objective? Many examples in this book come from an article written by Professor Wolynes entitled "Landscapes, funnels, glasses, and folding: from metaphor to software". Rather than trying to convince you, I will quote here the first few sentences of his article. "Of all intellectuals, scientists are the most distrustful of metaphors and images. This, of course, is our tacit acknowledgment of the power of these mental constructs, which shape the questions we ask and the methods we use to answer these questions."

---

[4] Reprinted excerpt with permission from Wolynes PG. (2001) Landscapes, Funnels, Glasses, and Folding: From Metaphor to Software, Proceedings of the American Philosophical Society **145**: 555–563

Time-based progression

Another instance of a non topic-based progression, the time-based progression is commonly found in the methodology section of any scientific paper.

> "The protein when it is **first made exists** in an extraordinarily large variety of shapes, resembling those accessible to a flexible strand of spaghetti. (2) The Brownian motion of the protein strand **will carry** it willy-nilly between various shapes, somehow **finally** getting it to settle down into a much less diverse family of shapes, which we call the "native structure" of the protein."[5]

In this time-based progression, the passage of time is expressed by changes in the verb tenses from the present to the future, and by the adverbs *first* and *finally*.

Although words such as *first, to start with, then, after, up to now, so far, traditionally, finally,* and *to finish* mark the start, the middle, or the end of a time step, time is often implicit. The reader scientist understands that the writer is following the logic of time when narrating the various steps of an experiment. Most often, the passage of time is established by changing the tense of a verb, from the past to the present, or from the present to the future.

Numerical Progression

The numerical progression is often used when describing a multi-step process. It is most clear when the number of steps is announced upfront, allowing the reader to know exactly how far along they are in the process. For example: "Minimizing losses during pulp extraction requires a delicate 6-step process from collection to treatment. Step 1: [...] Step 2... etc.

Announced Progression

Progression can be numerical, but it can also follow an order defined by the author (the elements of a list, for example). In the next sentence, the author announces two factors that contribute to the propagation of dengue fever before covering each one in turn.

> Two factors contribute to the rapid spread of dengue fever: air transportation, and densely populated areas.

Logical Progression

The progression can follow implicit logic, or explicit logic such as the cause and effect relationship announced by the words thus, therefore, because...

> The earnings report revealed a disastrous loss in gross income over the last two quarters. As a result, many employees, even those with seniority, feared for their jobs.

However, beware of logical transitions, as they may sometimes mask knowledge gaps, as in the following example:

> In December, the temperature in crowded Cambodian outdoor markets can sometimes drop a few degrees, therefore shopowners drink very little water.

Likely having no experience with Cambodian markets, this logical transition might perplex you. What connection is there between markets, the temperature, and water intake? And yet I can assure you that the sentence above is factual, and would indeed be considered logical to someone more knowledgeable of Cambodia.

This logical transition hides two facts. Firstly, the general knowledge that a drop in temperature can increase the pressure on the bladder, and thus require more frequent trips to the bathroom. And secondly, that bathrooms in Cambodian outdoor markets are rare and/or far, so going to the bathroom would require shop owners to abandom their stalls temporarily. This can be very inconvenient, so shopowners instead reduce their need to go to the bathroom at all... by drinking less water.

Now when you re-read the example above, you no longer have those logical gaps and can make full sense of the sentence. Remember to re-read not only for text fluidity, but for cognitive fluidity.

Progression through transition words

Progression is sometimes announced by transition words, such as *in addition, moreover, furthermore, and, also, besides*. These linking words are a topic of controversy among people who teach writing.

Some say they are just a convenient way to ignore progression; that they artificially establish a link where none exists. Unfortunately, they are often right. I recommend that when you see these transition words, you try to replace them with an implicit progression, such as a sequential step or a topic progression. If you cannot replace them, it may be that an explicit progression using these transition words is necessary, as in this conclusion paragraph:

> Our method determined the best terminal group for one specific metal-molecule coupling. In addition, we confirmed that Smith's formula for calculating molecular binding energy is less computationally intensive and more accurate than the frequently used formula proposed by Brown [8].

The two sentences are independent. *'In addition,'* here, emphasizes the extent of the contribution of the paper.

Pause in Progression

Progression is not a goal in itself. Sometimes the reader, despite the best efforts of the writer, gets lost because of insufficient knowledge or because too many distractions occurred while reading. To let the non-expert reader catch up, the writer needs to pause, in particular where the concepts take much concentration to understand, even for an expert! During that pause, the reader can consolidate the acquired knowledge with a writer's supplied summary or illustration. To announce that pause, use words such as *to summarize, briefly put,* and *for example* to name a few.

Troubleshooting Progression Problems

Unstructured paragraph

Sometimes it is possible to have every sentence in a paragraph be fluidly connected, and yet the whole paragraph remains difficult to follow. This is often because sentences progress from one topic to another through chain transitions without first exhausting all relevant mentions of that topic.

Given 3 sentences on topics A and C, and two sentences on topic B

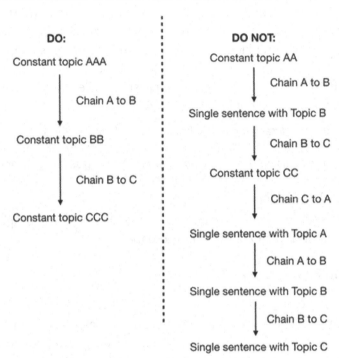

DO:

Constant topic AAA

Chain A to B

Constant topic BB

Chain B to C

Constant topic CCC

DO NOT:

Constant topic AA

Chain A to B

Single sentence with Topic B

Chain B to C

Constant topic CC

Chain C to A

Single sentence with Topic A

Chain A to B

Single sentence with Topic B

Chain B to C

Single sentence with Topic C

Interrupted topic

Sometimes, the progression breaks as in the example below. It stops for a sentence or two, and then is resumed. In such situations, readers rapidly lose their sense of direction. Somewhere, somehow, one or two links in the progression chain are broken, but where? How does one identify a broken link?

> (1) *After conducting microbiological studies on the cockroaches collected in our university dormitories, we found that their guts carried staphylococcus, members of the coliform bacilli, and other dangerous micro-organisms. (2) Since they regurgitate their food, their vomitus contaminates their body. (3) Therefore, the same microbes, plus molds and yeasts are found on the surface of their hairy legs, antennae, and wings. (4) To find such micro-organisms in their guts is not surprising as they are also present in the human and animal feces on which cockroaches feed.*

The topic of sentence 1 is about cockroaches, so links nicely to sentence 2 which reuses cockroaches as a topic. Sentence 2 and 3 are nicely linked through a logical connection "since" - > "Therefore". However, the fluidity breaks when attempting to move from sentence 3 to sentence 4. Can you identify why? Did anything feel familiar when reading sentence 4?

Sentence 4's topic concerns the micro-organisms that are found in cockroach guts. This also happens to be the stress of sentence 1, so a chain connection is possible here. If you restructure the paragraph to restore this connection the text becomes fully fluid:

> *(1) After conducting microbiological studies on the cockroaches collected in our university dormitories, we found that their guts carried staphylococcus, members of the coliform bacilli, and other dangerous micro-organisms. (4) To find such micro-organisms in their guts is not surprising as they are also present in the human and animal feces on which cockroaches feed (2) Since they regurgitate their food, their vomitus contaminates their body. (3) Therefore, the same microbes, plus molds and yeasts are found on the surface of their hairy legs, antennae, and wings.*

The next paragraph is about a tropical and sub-tropical disease called dengue fever. Compare the original version with the final version to discover how the text improved.

Original text

> *(1) The transmission of the dengue virus to a human occurs through the bite of an infected female Aedes mosquito. (2) In addition, the disease spreads rapidly in densely populated areas because of the lack of effective mosquito control methods, the increase in air travel, and poor sanitation in areas with a shortage of water. (3) The mosquito becomes infected when it feeds on a blood meal from a human carrier of the virus. (4) The virus multiplies inside the infected mosquito over three to five days and resides within its salivary gland.*

Follow these steps to analyze the original text.

- Identify the author's intention, and the main points of the paragraph.
- Identify closely related sentences by the words they share.
- Identify potential sentence topics — usually the words repeated from one sentence to the next.
- Restructure the text to achieve progression and set desired expectations around one topic.

Solution.

- Author's intention and key points

The author seems to be making two points in this paragraph. Firstly, an explanation of the propagation cycle of the dengue virus (sentences 1, 3, and 4); and secondly, the conditions for rapid disease spread (sentence 2).

- Topic and progression scheme

There are at least two possible topics: the virus (repeated three times) and the mosquito (repeated four times). Either one can be the constant topic in all sentences. Regarding sentence order, since the paragraph (not shown here) that follows our sample paragraph describes what the community can do to stop the propagation of dengue fever, the last sentence must be similar to sentence (2).

- Text restructure taking into account the reader's prior knowledge and anticipating what the reader may not know

Having monitored how thousands of scientists rewrite this paragraph in our workshops, I would like to warn you about potential time-wasting pitfalls. Some of you may think that reordering the sentences to fix the lack of progression is enough. You clearly see that the second sentence is out-of-place and move sentence 2 to sentence 4.

> *(1) The transmission of the dengue virus to a human occurs through the bite of an infected female Aedes mosquito. (3) The mosquito becomes infected when it feeds on a blood meal from a human carrier of the virus. (4) The virus multiplies inside the infected mosquito over three to five days and resides within its salivary gland. (2) In addition, the disease spreads rapidly in densely populated areas because of the lack of effective mosquito control methods, the increase in air travel, and poor sanitation in areas with a shortage of water.*

However, this creates a disconnection between part 1 of the paragraph (the transmission cycle) and part 2 (spread of the disease). Why? After moving sentence 2 to the end of the paragraph, sentence 4 now becomes sentence 3 which ends with "and resides within its salivary gland." Moving from the *salivary gland* of a mosquito to *the spread of the disease* is a conceptual jump too great for the reader. Therefore, do not describe the cycle using the sequence Mosquito infects Human — Human infects Mosquito, but instead choose the sequence Human infects Mosquito — Mosquito infects Human. That way, the salivary gland sentence will be the second sentence in the paragraph, in the middle of the sequence. In that position, it does not create any problem.

The sentence that ends the paragraph has three problems.

In addition, the disease spreads rapidly in densely populated areas because of the lack of effective mosquito control methods, the increase in air travel, and poor sanitation in areas with a shortage of water.

1. The bogus transition word "in addition" must go. But even when it is removed, the sentence starts with the wrong topic: the disease. Either you write an additional sentence with words extracted from the last sentence to act as glue between the two parts (cycle and spread), or you find in that last sentence old information that can be brought upfront and used as the sentence topic.

2. Examine the information given in that sentence. Is all of it useful? Remember that the next paragraph following the one you rewrite is about <u>community</u> action to stop the spread of the disease.

3. You are probably new to the topic of dengue fever. Yet, you know that mosquitoes need water to breed their larvae. Knowing this, does everything in that sentence make sense? Is there a knowledge gap that needs to be bridged?

Final text (version one — Virus is the constant topic)

The dengue virus *from a human carrier enters the female Aedes mosquito via the infected human blood she feeds on.* ***The virus*** *then multiplies inside the mosquito's salivary gland over three to five days.* ***It*** *is transmitted back into another human through the saliva injected by the infected mosquito when she bites.* ***The virus*** *<u>spreads</u> rapidly in areas where large numbers of humans and mosquitoes cohabit. <u>This spread</u> is accelerated with human travel (air travel particularly), ineffective mosquito control methods, and poor sanitation in areas with water shortages.*

Progression is built around a constant topic. The progression is also a time-based progression (the transmission cycle) and a logical progression (amplification: from limited to extended, from specific to general). The fourth sentence acts as a bridge between the two parts. It is built by removing an element from the last sentence of the original version — 'densely populated areas' — and developing it as a full sentence.

This paragraph is understandable by experts who have lived in tropical countries, but not by "the rest of us". Some say: I know that mosquitoes breed in water, so if there is a shortage of water, how do mosquitoes reproduce? The lack of knowledge leads to a lack of understanding even though the text is clear.

Final text (version two — Mosquito is the constant topic)

*The female Aedes **mosquito** feeds on the infected blood of a human carrier of the dengue virus. Inside the **mosquito** salivary gland, the virus multiplies over a period of three to five days. Subsequently, when the infected **mosquito** bites, its saliva carries the virus back into another human. Ineffective **mosquito** control methods cause the dengue virus to spread rapidly, particularly in densely populated areas at the end of the rainy season due to the sporadic rains. At that time, **mosquitoes** enjoy abundant breeding sites such as puddles of stagnant water and large opened vats storing rainwater for the dry season.*

In the second version, the mosquito is the constant topic in all sentences. The connection between the end of the transmission cycle and the spread of the disease is established simply by bringing upfront in the rewritten sentence an element — *'ineffective mosquito control'* — that was in the middle of the last sentence in the original version. Some information from the original sentence was removed, and extra details were added to make this paragraph clearer and to prepare the transition to the next paragraph on community action. The disappearance of *air travel* is intentional since air travel is not directly relevant to the villagers who are part of the community; and so is the removal of *'water shortages'*, which is counterintuitive since mosquitos need water to breed. The additional information clarifies the link between the shortage of water and the spread of the disease.

Final text (version three — sequence progression and Cycle topic)

The reproduction of the dengue virus relies on a three-step **cycle**. **First**, the virus enters the female Aedes mosquito when the mosquito feeds on the infected blood of a human host. Once inside the mosquito salivary gland, the virus **then** multiplies over a period of three to five days. The **cycle** is **complete when** the virus returns to another human through the saliva injected by the infected mosquito when it bites. This **cycle** is repeated rapidly (sometimes at epidemic speed) in densely populated areas when people travel, and when mosquito control is ineffective, particularly in areas with water shortages where mosquitoes breed in open air water storage containers.

In this case, the progression is sequential: three steps. The reader is counting. When the sequential progression ends, the constant topic "Cycle" returns at the end of the paragraph.

Final text (version four — greater knowledge gap — children with some prior knowledge on mosquitos)

People sick with Dengue fever carry the dengue virus inside their blood. When that blood is sucked by a female mosquito called Aedes Aegypti — that's her family name — the blood carries the virus inside of her. It does not harm her, it simply invades her body. Fifteen days after it got in, it arrives inside her saliva and stays there to multiply, and multiply, and multiply for 3 to 5 days. So when the female mosquito bites someone else, her saliva carries a whole army of dengue viruses into the blood, enough to make that person sick. Where many mosquitos and people live together, dengue fever spreads very fast. It spreads even faster when people travel by plane, or when people are careless and let the Aedes Aegypti mosquitos lay thousands of eggs on the side of open water containers like old tires or flower pot plates.

Children require less sophisticated vocabulary. *Carrier, infected, densely populated, control method,* these are all adult words. They need spoken language, not written language. They need intermediary explanations, examples, numbers, personification, and drama.

One final word of caution: Do not attempt to "fix" progression problems in a paragraph without taking into account the topic of the next paragraph. Progression applies between paragraphs as much as it applies between the sentences of a paragraph. Progression problems are not always fixed by moving sentences around. In many cases, an unclear text needs complete restructuring. To restructure, understand the author's intention, and identify the key points of the argument presented and their underlying logical connections.

Take ten consecutive sentences from the discussion of your paper. Identify the topic and the stress of each sentence. Can you identify a progression scheme? Are some sentences totally disconnected from their predecessors? Are sentences artificially connected by a transition word hiding a progression problem? Rewrite these sentences to restore normal progression.

Chapter 11

Detect Sentence Fluidity Problems

 Here is a very important exercise you don't want to miss because it allows you to evaluate how well you set expectations. Follow it step by step. It will only take you ten minutes.

You can also do this using SWAN – the Scientific Writing Assistant, a software tool developed in conjuction with the writing class. The tool can be found at http://cs.joensuu.fi/swan/. There are help videos on the website for general use, but here we'd like you to focus on the manual fluidity assessment function.

Step 1:
Choose a paper you wrote, or if you have never written a paper, choose any paper in your field. Take a pencil. You will need it to underline words in each sentence, and to draw one of the smiley faces at the beginning of each sentence.
☺ ☺ ☹

Step 2:
Select the first eleven sentences from the introduction of the paper.

Step 3:
Set your eyes on the first sentence of the introduction.

Step 4:
Read the sentence and block eye access to the next sentence with a piece of paper. After you understand the sentence, underline the words that clue you on the topic of the next sentence. If you have an expectation, go directly to Step 6 (skip the next step).

Step 5:

You have no expectation for what could come next. Move to the right the paper hiding the next sentence until you get to the first verb of the sentence, and go no further. If you now have an expectation, move to step 6, but if you still have no expectation, draw a ☺ at the beginning of the sentence you examined (not the one you partially uncovered). Go directly to Step 7 (skip the next step).

Step 6:

Test your expectation to see if you guessed right. Look at the complete sentence you were hiding. If it fulfills your expectations, draw a ☺ at the beginning of the sentence you examined (not the one you just read to check whether your expectation was right), but if it breaks your expectations, draw a ☹ instead.

Step 7:

If you have finished you should have a smiley at the beginning of the first 10 sentences; go to Step 8. If not, go back to step 4 and continue with the next sentence.

Step 8:

Compute your score. For each ☺ give yourself a pat on the back and add three points to your total score. For each ☺ subtract one point, and for each ☹ subtract two points. If the score is twenty or more, this writing is brilliant. If the score is less than twelve, consider rewriting the ☺ and ☹ sentences to better control the expectations of the reader.

Here is an example:

> The sky was blue. (No idea, let me look up to the next verb)
> The sky was blue. The cat was (still no idea. I'll put ☺ in front of the first sentence)
> ☺ The sky was blue. The cat was trying to <u>catch a bee</u> on a flower. (I'll bet the cat will be stung by the bee!)
> The sky was blue. The cat was trying to catch a bee on a flower. The bee stung its paw. (Yeah! here is a ☺ I was right)
> ☺ The sky was blue. ☺ The cat was trying to catch a bee on a flower. The bee <u>stung its paw</u>. (The cat limped away)
> ☺ The sky was blue. ☺ The cat was trying to catch a bee on a flower. The bee stung its paw. The gardener dug a hole. (☹ Wrong!)

☹ *The sky was blue.* ☺ *The cat was trying to catch a bee on a flower.* ☹ *The bee stung its paw. The gardener dug a hole.*
Total score: [−1, +3, −2] = 0!

Now that you are finished with the exercise, wouldn't you like to find out the cause of these faces — in particular the neutral and sad faces? You could do it by yourself, going over each sentence to determine what created your expectations and how the writer violated them. You would then become aware of the causes of fluidity problems. In fact, why don't you do it now!

Many scientists did this exercise during the writing class. They identified the common problems. So, when you are done with the exercise, read below to check whether the problems you found are also described here.

Reasons for No Expectations ☺

Rear-mirror sentence

"The above framework is compliant with traditional digital signature system structures."[1]

In this sentence, the author adds one final detail that ends a rather complete presentation of the framework. The sentence looks backward, not forward. It sets no expectation.

Descriptive sentence

"The coordinates of each point in a uniform dataset are generated randomly in [0, 10000], whereas, for a Zipf dataset, the coordinates follow a Zipf distribution skewed towards 0 (with a skew coefficient 0.8)".[2]

Sentences that state facts or data create little expectation.

Paragraph break

By definition, the paragraph is a unit of text with a single theme or purpose. When the theme or purpose changes, the paragraph

---

[1] Qibin Sun and Shih-Fu Chang, A Robust and Secure Media Signature Scheme for JPEG Images, Journal of VLSI Signal Processing, Special Issue on MMSP2002, pp. 306–317, Vol. 41, No. 3, Nov., 2005.
[2] Papadias Dimitris, Tao Yufei, Lian Xiang, Xiao Xiaokui The VLDB Journal: the international journal on very large data bases, July 2007, v.16, no. 3, pp. 293–316.

changes. But how are these two themes connected? Do you let readers find out by themselves? If you do, is the connection obvious even to a non-expert? It is best to make that connection explicit in one of two ways: use the last sentence of the paragraph to introduce the theme of the next paragraph, or use the first sentence of the new paragraph to introduce the new theme and show how it relates to the previous paragraph.

In this next example, the connection is not explicit. The paragraph states that there are well-known algorithms and others not as well-known (we presume). But we do not know what to expect next.

"Many clustering algorithms are proposed to group co-expressed genes into clusters. Some well-known examples include hierarchical clustering [1, 2], K-mean clustering [3], and self-organizing maps [4, 5]. Bayesian networks [6] and graph theoretic approaches [7–9], model-based methods [10–12], and fuzzy clustering [13] provide additional methods toward clustering.

Various measurements have been employed to score the similarities between pairs of gene expressions."

Enumeration with no predictable item length

If each item of a numbered enumeration consists of one sentence, the reader very rapidly discovers the pattern and expects the situation to continue. But if each item is covered by a variable number of sentences, the reader does not know what to expect after the end of each sentence: a new item, or the next sentence on the same item.

In science, two factors contribute to long sentences: precision and intellectual honesty. Adding precision to a noun through a modifier lengthens the sentence. For example "an R400-7 fiber optic reflection probe with 6 illumination fibers and one read fiber" is significantly more precise than "a fiber optic probe". Such precision may be necessary to convince the reader of the quality of the fluorescence measurement. Intellectual honesty lengthens a sentence through the addition of detailed qualifiers and provisos.

In this example, the reader expected the writer to move to the "intellectual honesty" factor, right after the "for example" sentence. But instead, the writer adds a sentence on the probable need for precision. The reader, whose expectations are now broken, resets expectations, but no longer knows what to expect after the comment. This situation could have been avoided by removing the sentence. It is weak ('may be' expresses uncertainty), and non-essential.

Vortex of chained explanatory details creating an expectation void

It is a case of "one thing leads to another", but with no end in sight. The reader follows but does not know what to expect next. A good example has already been given. It is repeated here.

> For the next three days, the thirty embryonic cells proliferate in the culture **dish**. The dish, made of plastic, has its inner surface **coated** with mouse cells that through treatment have lost the ability to divide, but not their ability to provide nutrients. The reason for such a special **coating** is to provide an adhesive surface for the embryonic cells. After proliferation, the embryonic cells are collected and put into new culture dishes, a process called "replating".

The first sentence establishes cell proliferation as the paragraph topic. The next two sentences take us down a series of chained details: from embryonic cells to culture dish, from culture dish to dish coating, and from dish coating to adhesive surface. The reader does not know what to expect next. The return to the proliferation topic is not surprising, but it is not expected.

Vague, general statement

> With the rapid development of computer communication and Internet, the distribution of digital images is pervasive.

The next sentence is about watermarking. Who would have guessed! The statement is general and typical of the first vacuous sentences of an introduction. From such a vague statement, no expectations can be derived.

Long sentence diluting expectations across many topics

> "In this paper, we focus on extending the implementation of the Run Time Infrastructure (RTI) to relax the time synchronization among federates, particularly focusing on RTIs that support the conservative simulation protocol for their time management service, for example the DMSO's RTI."[3]

What is the next sentence to be about: RTI, federates, relaxing time synchronization, the conservative simulation protocol, the time management service of the DMSO's RTI? The reader has no clue what to expect because there are too many options in this long sentence filled with details. The next sentence is about the method used by each federate to regulate time. Who would have guessed? I thought that the writer would explain why there was a need to relax time synchronization. I would have given that sentence a ☺ but the experts in the field have given it a ☺ (**Expectations vary from one reader to the next, depending on their background knowledge.**)

Reasons for Betrayed Expectations ☹

Unannounced topic change and ambiguous pronoun

> *The images in our set are normalized to have a consistent gray level when output. The coordinates of the eyes are automatically detected and set to a fixed position. They are then resampled to a given size. After normalization...*

The topic of the first sentence suggests two possible candidates for the next sentence:

1. The benefit of having a consistent gray level
2. Additional normalization step on the images, such as image contrast

The "eye" topic of sentence 2 does not fit expectation 1 or 2. It could have, had the writer specified that the images were face images. Only after readers finish reading sentence 2, do they realize that it ties in with candidate 2 — the additional normalization step. Based on this, readers expect two new candidates for the next sentence:

1. Yet another normalization step
2. A reason for fixing the eye coordinates, or the location of the fixed position

---

[3] Boon Ping Chan, Junhu Wei, Xiaoguang Wang, (2003) Synchron, Synchronization and management of shared state in hla-based distributed simulation, Proceedings of the Winter Simulation Conference,S. Chick, P. J. Sánchez, D. Ferrin, and D. J. Morrice, eds.

The pronoun *'they'* starts the next sentence. Readers think the pronoun refers to *'the coordinates of the eyes'*, and they expect a reason for fixing these coordinates. But they reach an impasse when the verb "resampled" arrives because coordinates are unlikely to be resampled. Readers briefly stop and re-read to discover that *'they'* refers to the images mentioned in sentence 1. The pronoun was ambiguous. The third sentence was simply another normalization step (expectation 1).

Here is a possible rewrite.

All face images are normalized (1) to display consistent gray levels, (2) to automatically place the eyes at the same location in each image, and finally (3) to set each image to a 128 × 128 size through resampling. After normalization, ...

Text precision is increased. The sentence is long, but precise and easy to follow. The enumeration and the parallel syntax set the expectations that we are moving from one normalization step to the next.

Adjectival or adverbial claim not followed by evidence

Adjectives or adverbs are words that set expectations in three ways:

- They make a claim that requires validation, as in *"The world is flat"*.
- They state an existing (often well-known) situation judged unsatisfactory that requires correction, as is *"Airplane engine maintenance is very labor-intensive"*. In this case, *very labor-intensive* sets the expectation for its opposite, less labor-intensive. Note the role of *very* as a judgmental word to enhance the expectation for a more satisfactory solution.
- They state a situation in a negative or a pejorative light that requires the need for a change towards the positive or the ameliorative as in *"trapping is unimportant at high temperatures."* The sentence sets the expectations that trapping may be important at low temperatures.

*Drop test simulations that rely on a fine mesh of 3D finite elements are very **time-consuming**. They are required to save the time and cost involved in actual physical tests conducted at board level. We* *propose a simplified model that considers the board as a beam structure...*

The adjective *time-consuming* makes a claim associated to the Finite Element Method. Readers in that field are already aware that fine mesh 3D modeling is very time-consuming. They expect the writer to present faster methods in the next sentence. It comes... in the third sentence. You may object that the writer uses a good progression scheme, the constant topic progression: *'drop test simulations'*, and the pronoun *'They,'* although the pronoun is ambiguous because it could also replace *finite elements*. To answer the objection, it is worth emphasizing that **expectations import more than progression**. To meet the expectations of the reader, these sentences must be rewritten. Here is a possible rewrite.

Simulations based on the analysis of a fine 3D mesh of finite elements save the cost, but not the time involved in conducting physical board level drop tests. Saving time is possible if the model is simplified. We propose a model that simplifies the board by making it a beam structure...

In this rewrite, chain progression is set, and expectations are met.

Unclear answers to clear questions

Beyond which threshold value should the degree of asymmetry of the brain be considered abnormal? The answer depends on what kind of brain information is required. When one studies pathological abnormality, false positives and false negatives should be kept low. The threshold value differentiating normal from abnormal asymmetry could be estimated from patient data, but how sensitive or how specific that value would be, is unclear.

The starting question sets an expectation for a precise numerical value. The *it-depends* answer is unclear and disappoints. Indeed, the second sentence acts as a second indirect question: What is the brain information required? Again, the answer disappoints, because the writer gives an unexpected answer that reveals a concern for the statistical quality of the data. The reader could not possibly have guessed that, and is therefore surprised by the false positives/false negatives sentence. At this point, the reader usually gives up and, having been surprised twice in a row, takes a neutral attitude. The writer then tells the reader that a threshold value is of little use if it is not sensitive or specific. Here is a possible rewrite:

Assuming that the degree of brain asymmetry is a reliable indicator of brain pathology, can a diagnostic of abnormality be based on an asymmetry threshold value? To answer that question, it will be necessary to determine whether such a value derived from patient data has enough sensitivity and specificity.

Unjustified choice

"The methods for cursor control [Brain Computer Interface] come under two categories, regression [references] and classification [more references]. Each of them has its merits. In our study we adopt the classification method,..."[4] (the text that follows describes the method, but does not give the reason for the choice).

The writer presents the choice he faced: regression or classification. Both have their own advantages. The reader has two possible expectations:

- The writer will give the advantages and disadvantages of each. Obviously, the writer did not intend to do that. It would be better to remove the sentence *"Each of them has its merits"* because it creates the wrong expectations.
- The writer will choose one category of methods and justify it.

Which disadvantage, or which advantage led the writer to prefer the classification method? The writer does not say. Note that even if the second sentence *"Each of them has its merits"* is removed, the reader will still want to know why the writer chose the classification methods over the regression methods. Whenever you announce and make a choice, the reader wants to know why, even if the choice is arbitrary.

---

[4] Zhu X, Guan C, Wu J, Cheng Y, Wang Y. (2005) Bayesian Method for Continuous Cursor Control in EEG-Based Brain-Computer Interface. *Conf Proc IEEE Eng Med Biol Soc* **7**: 7052–7055.

The methods for cursor control [Brain Computer Interface] come under two categories, regression [references] and classification [more references]. We decided to adopt the classification method because...

Broken repeated patterns and paraphrase

The emergence of Hidden Markov Models (HMM) in the 1970s allowed tremendous progress in speech recognition. HMMs are still the de-facto method for speech recognition today. However, some argue that HMMs are not a panacea. At the same time, today's speech recognition systems are far smarter than those in the earlier days.

The first sentence establishes the paragraph topic: HMMs. It places it in a good light. The reader expects the praise to continue, and it does in sentence 2. The reader is given two expectations for sentence 3: (a) examples of the current use of HMMs in speech recognition, (b) or the reasons for the continued use of HMMs (maybe recent improvements to the basic HMM method).

In sentence 3, the writer instead surprises the reader by announcing the limitations. Expectations reset, the reader now expects the writer to name a problem where HMMs do not excel. Unfortunately, in sentence 4, the writer seems to change topic. *"At the same time"* is a phrase similar to *"in addition"*: it masks the lack of a proper transition.

When breaking a pattern, it is better to announce the break in the same sentence as shown below.

Although HMMs are still the de-facto method for speech recognition today, they are not a panacea. As speech recognition applications increase in sophistication (automatic language recognition, speaker authentication), HMMs need to blend harmoniously with other statistical methods, in real-time.

Lack of knowledge (or unexplained word) and synonyms

Blood vessels carry blood cells as well as platelets, which are not as numerous as blood cells (ratio of 1 to 20). Platelets, besides being a source of growth factors, are also directly involved in homeostasis by aggregating to form a platelet plug that stops the bleeding. A thrombocyte count is usually included in a blood test as it is a means to determine diseases such as leukemia.

The third sentence blocks the non-expert reader who does not know that thrombocyte is synonymous with platelet. In this case, the writer did not wish to repeat platelet because it had already been mentioned twice in the previous sentence. Filling a paper with synonyms increases the knowledge gap and increases the memory load. As a rule, settle on one keyword (the simpler one) when two synonyms are available, and use it consistently in your paper.

A platelet count is usually included in a blood test as it is a means to determine diseases such as leukemia.

But if you wish to introduce a new keyword, define it in a just-in-time fashion as shown below.

A platelet count (also known as a thrombocyte count) is usually included in a blood test as it is a means to determine diseases such as leukemia.

Chapter 12

Control Reading Energy Consumption

Réponse hémodynamique

I learned something truly fascinating by reading one of researcher Peter Hagoort's neuroscience papers. It described what happens in our brain when, during reading, it encounters strange things such as "the car stopped at the casserole traffic light". Something similar happened to me when I stumbled over the word 'hemodynamic' in the article. Google took me to the website fr.wikipedia.org/wiki/Réponse_hémodynamique, and things became very interesting. I discovered that when reading becomes difficult, the body sends a little more blood (i.e. glucose and oxygen) to the brain. It does not take blood from one part of the brain to send it to another part so as to keep energy consumption constant; it simply increases the flow rate. Following the trail like a bloodhound, I discovered a French article written by André Syrota, director of the life science division at the Atomic Energy Commission, indicating that our brain's additional work could consume the equivalent of "147 joules per minute of thought."[1]

How tired will your readers be at the end of their reading journey? How well did you manage their time and energy? As George Gopen points out, reading consumes energy.[2] Reading scientific articles consumes A LOT MORE ENERGY. Therefore, how do you reduce the reading energy bill, and how do you give your reader the assurance that plenty of energy refueling stations will be available along the long and winding road of your text?

---

[1] http://histsciences.univ-paris1.fr/i-corpus-evenement/FabriquedelaPensee/affiche-III-8.php

[2] George D. Gopen. 2004. "Expectations: teaching writing from the reader's perspective". Pearson Longman, p. 10.

The Energy Bill

Let, E_T, be the total energy required by the brain to process one sentence. E_T is the sum of two energies:

1) The syntactic energy E_{SYN} spent on analyzing sentence structure
2) The semantic energy E_{SEM} spent on connecting the sentence to the others that came before it, and on making sense of the sentence based on the meaning of its words.

Gopen considers these two energies to be in a "zero-sum relationship".[3] This means that if E_{SYN} becomes large, it will be at the expense of E_{SEM}: the more energy is spent on the analysis of the syntax of a sentence, the less energy will be left to understand the meaning of the sentence.

$$E_T = E_{SYN} + E_{SEM}$$

E_T is finite and allocated by the brain to the reading task. Similarly to our lungs, which give us the oxygen we need one breath at a time, the brain has enough energy to read one sentence at a time. E_T cannot increase beyond a certain limit fixed by physiological mechanisms: to increase the blood flow rate takes a few seconds and the size of the blood vessels in the brain (although extensible) is limited.

Your job, as writer, is to make sure that $E_{SYN} + E_{SEM} < E_{Tmax}$ at all times by minimizing both the syntactic and the semantic energy required to read.

What would consume excessive syntactic energy, E_{SYN}?

- Anything ambiguous or unclear — a pronoun referring to an unclear noun, a convoluted modified noun, an ambiguous preposition
- Spelling or light grammar mistakes — missing definite article *the*, wrong preposition such as *in* instead of *on*, misused verb such as *adopt* instead of *adapt*

---

[3] George D. Gopen. 2004. "Expectations: teaching writing from the reader's perspective". Pearson Longman, p. 11.

- Incomplete sentences, i.e. missing verb
- Anything taxing on the memory — long sentences (usually written in the passive voice), formulas, sentences with multiple caveats, provisos and qualifiers, sentences with deeply nested subordinates
- Grammatical structures from a foreign language applied to English without change
- Missing or erroneous punctuation

What would consume little syntactic energy, E_{SYN}?

- Short sentences with simple syntax: subject, verb, object

   New ideas disrupt the logical flow of sentences.

- Sentences with a predictable pattern established with words such as *although, because, however,* or *the more...the less.*

   The more energy is spent to analyze the syntax of a sentence, the less energy is left to understand what the sentence means.

- Sentences with subject close to verb, and verb close to object

   Motivation allocates the total energy, E_T, to the reading task.

- Sentences with good punctuation

   The reader has three choices: give up reading, read the same sentence again, or read what comes next.

What would consume great semantic energy, E_{SEM}?

- Unknown words, acronyms, and abbreviations
- Absence of context to derive meaning
- Lack of prior knowledge to understand, or to aid understanding
- Lack of examples or visuals to make the concept clear
- Overly detailed or incomplete visuals
- Reader forgetful of content previously read
- Reader in disagreement with statement, method, or result
- Very abstract sentences (formulas)
- Sentences out of sync with reader expectations

What would consume little semantic energy, E_{SEM}?

- Sentence with a well-established context

    Total reading energy for a given sentence, E_T, is the sum of two elements: the syntactic energy, E_{SYN}, spent on analyzing its syntax, and the semantic energy, E_{SEM}, spent making sense of the just analyzed sentenced.

$$E_T = E_{SYN} + E_{SEM}$$

- Reader familiar with the topic or the idea

    The songbird flew back to the nest to sit on three little eggs, two of its own, the third one from a cuckoo.

- Sentence that explains the previous sentence

    Therefore if E_{SYN} becomes large, it will be at the expense of E_{SEM}. The more energy is spent to analyze the syntax of a sentence, the less energy will be left to understand the meaning of the sentence.

- Sentence that prepares the ground (through progression or setting of context)

    Sub-clauses that pull reading forward often follow a predictable pattern; they start with a preposition such as "although," "because," "however," or "if".

- Short sentences (with known vocabulary)

    It does not. The reader is surprised.

What would get the reader into trouble?

Energy shortages occur when $E_{SYN} + E_{SEM} > E_{Tmax}$

- E_{SYN} is unexpectedly large. As a result, what remains of E_{SEM} is insufficient to extract the complete meaning of the sentence.
- E_{SYN} is normal; but a new word, acronym, abbreviation, apparent contradiction, or concept requires additional brain effort (saturated memory, or failure to find associative link with known data). The reader runs out of E_{SEM}. The semantic energy gas tank is empty before the sentence is fully understood.

When this happens, the reader can make one of three choices: give up reading, read the same sentence again, or read what comes next, hoping to understand later.

Giving up reading is tragic. It is a consequence of repetitive and successive breakdowns in understanding. The text becomes increasingly obscure, and the reader finally gives up reading.

Reading the same sentence again happens if motivation is high. The reader is determined to understand because much understanding is expected from the difficult sentence. Rereading after mastering a difficult syntax consumes no syntactic energy because the sentence syntax is now familiar, and the reader spends all of his or her energy on understanding only.

$$E_{SYN} = 0, \text{ and therefore } E_T = E_{SEM}.$$

The simile — reading as consuming brain energy — is in line with what Science observes. The brain hard at work consumes more energy.

Punctuation: An Energy Refueling Station

The period, a full-stop to refuel

When the period arrives, the reader pauses and refills his or her energy tank before reading the next sentence. It gives the reader a chance to conclude, absorb, consolidate the knowledge just acquired and anticipate what comes next (from expectations or progression).

The semicolon, a fuel stop for topping a half-full tank

Surprisingly, searching for a semicolon in a scientific paper is often rewarded by the infamous beep that says: "None found, can I search for anything else?" Periods, colons, and commas seem to be the only punctuation used by scientists. Scientists, by nature logical, should be fond of semicolons, not only to strengthen their arguments, but also to make their text less ambiguous and to carry the context forward at little cost.

Semicolons are close cousins to the period. They stand at a place of semantic closure, like the period. They end and start a clause (part of a sentence with a subject and verb). Unlike the period, their role is to unite, join, or relate; while the role of the period is to separate. The clauses on each side of a semicolon are often compared, contrasted, or opposed. Often, the first clause in the sentence makes a point, and the clause (or clauses) after the semicolon refines, details, or completes the point. Semicolons are often found close to conjunctions and conjunctive adverbs, such as 'but,' 'consequently,' 'however,' 'therefore,' 'thus,' or 'nonetheless'.

> *The calculated data and the observed data were closely related; however, the observed data lagged when concentration dropped.*

The first clause in the above sentence provides context to the whole sentence. It shares its subject with the clause that follows the semicolon. These two clauses are also closely related semantically, much more so than two sentences separated by a period. Therefore, since the context does not vary within the sentence, reading is faster and easier.

A semicolon has more than one use. When a sentence needs to be long to keep together a list of sentences, the semicolon does the job magnificently.

> *Information with visual impact requires creativity, graphic skill, and time. Because most of these are in short supply, software producers provide creativity, skill, and time-saving tools: statistical packages that crank out tables, graphs and cheesy charts in a few mouse clicks; digital cameras that in one click capture poorly lit photos of test bench equipment replete with noodle wires (I suppose the more awful they look, the more authentic they are); and screen capture programs that effortlessly lasso your workstation screens to corral them for your paper.*

The three ":" "!" "?" fuel stops and the comma

Other punctuation marks provide an opportunity to refuel: the colon, the question mark, and the unscientific exclamation mark (I wonder if Archimedes would have damaged his reputation as a scientist had he ended his "Eureka" statement with an exclamation mark). The question mark, the most underused punctuation mark, causes the reader to pause and think while also setting clear expectation for an answer. The colon introduces, explains, elaborates, recaps, and lists. Unlike the semicolon, it can be followed by a phrase that lacks a verb. Like the semicolon, it is preceded by a whole clause (not one truncated as in the next example).

And the results are:

In a correct sentence, the main clause is not truncated.

And the results are the following:

Colons are much liked by readers: they announce clarification or detail. Colons are also the allies of writers. They help introduce justification after a statement.

Commas help to disambiguate meaning, pause for effect, or mark the start and end of clauses. But, for all their qualities, there is one that commas cannot claim: semantic closure. Readers cannot stop at a comma and decide that the rest of the sentence can be understood without reading further.

In this chapter, you have been given many tools to reduce the reading energy bill. Imagine your writing as a piece of cloth, and the brain of the reader as an iron. If your writing has the smoothness of silk, the iron can be set to the lowest temperature setting. If it has the roughness of overly dry cotton, not only will the iron have to be set to the highest temperature setting, but you will also put the reader under pressure and demand steam to iron out the ugly creases in your prose. Either **you** spend the time and energy, or **the reader** does.

Ask a reader to read your paper and to highlight in red the sentences not clearly understood (high semantic energy), and in yellow, the sentences that slowed down reading because of a difficult syntax (high syntactic energy). Then correct accordingly.

Some punctuation helps reduce reading energy. Search for ':' and ';'. Do you have enough of them? If not, look for opportunities to use them, particularly in long sentences.

Part 2

Paper Structure and Purpose

Each stage in the construction of a house contributes to its overall quality. Similarly, each part in an article contributes to the quality of the whole, from the abstract (the architect blueprint) and the structure (the foundations), to the introduction (the flight of steps and the porch), the visuals (the light-providing windows), and finally the conclusions (the house key).

The art of construction is acquired through a long apprenticeship. You may be attracted by the time-saving expedient prefab or by the imitation of other constructions of uncertain architectural quality (the paper your supervisor handed out to you as a template for your first paper). Beware of shortcuts. A hastily assembled paper, once analyzed, often reveals major cracks and faults: its shapeless structure is like baggy jeans that would fit just about any frame; its graphics and other visuals have a mass-produced look and feel.

When your house is built, when your paper is finally published, how will the readers feel after visiting your house? In your lobby, you have left a guest book, in which your visitors can enter their remarks. Here are two entries. Which one would please you the most? And what kind of scientific paper would you associate such remarks with?

> To Whom It May Concern,
>
> *Concrete slabs and parched grass don't make much of a garden. I really don't know how you dare call that plywood platform a porch. It is laughable.*
>
> *The inside of the house is dark and gloomy. You need more windows. And by the way, I did not take anything from the fridge. There was nothing in it, and it wasn't even plugged in. I could not find the things you claimed were in the house. I looked everywhere, but everything was in such a mess, I gave up. There were large cracks in some walls (not line cracks). Your house is not safe. And while we are on the subject of walls, why did you paint them all a dull gray? Anyway, I'm out of here. Thanks for the spare key, but I left it behind. I don't intend to come back.*

> Dear House Owner,
>
> *Your house is such a delight. I wish I had one like that. Your flower garden and your large porch are so inviting. Your indoors are exquisitely designed. Everything is at the right place. I immediately found what I was looking for. But I did not expect the basket of fresh fruit and the cool complementary drinks in the fridge. How lovely! I love your large bay windows. They make the house look so bright and colorful. There is not one dark corner!*
>
> *Thank you for your spare key. I also took your business card to advertise this lovely house of yours, which makes people feel so welcome.*
>
> *With deepest thanks,*
> *Professor Higgins*

What makes a good house depends on its design, of course, but also on the attitude of the house owner towards his guests. A good writer is one who cares for his readers; someone who anticipates their needs. **Good writing is not just a matter of writing skills; it is also a matter of attitude towards the readers.**

To construct a satisfactory paper, one must understand the role each part plays for the reader and the writer. And to assess the quality of each part, one must establish evaluation criteria. The next chapters fulfill these objectives. Numerous examples are given to analyze and to help you distinguish good writing from bad writing. By the end of the book, you will be ready to write your paper.

First Impression

Today, as the city's bowels demonstrate their usual constipation, the pouring rain adds a somewhat slimy aspect to the slow procession of traffic. Professor Leontief does not like arriving late at the lab. He hangs his dripping umbrella over the edge of his desk, at its designated spot above the trashcan, and he gently awakens his sleepy computer with some soothing words "Come on, you hunk of metal and silicon oxide, wake up."

He checks his email. The third one is from a scientific journal where he helps out as a reviewer. "Dear Professor Leontief, last month you kindly accepted to review the following paper and submit your comments by....". He looks at his calendar and realizes that the deadline is only 2 days away. A cold chill runs up his spine. He hasn't even started. So much to do with so little time! Yet, he cannot postpone his response. Being a resourceful man, he makes a couple of phone calls and reorganizes his work schedule so as to free an immediately available 2-hour slot.

He pours himself a large mug of coffee and extracts the article from the pile of documents pending attention. He goes straight to the reference section on the last page to see if his own articles are mentioned. He grins with pleasure. As he counts the pages, he looks at text density. It shouldn't take too long. He smiles again. He then returns to the first page to read the abstract. Once read, he flips the pages forward slowly, taking the time to analyze a few visuals, and then jumps to the conclusions that he reads with great care.

He stretches his shoulders and takes a glance at his watch. Twenty minutes have passed since he started reading. By now, he has built a more or less definitive first impression on the paper. Even though the article is of moderate length, it is too long for such a small contribution. A letter would have been a more appropriate format than a full-fledged paper. He will have to inform the writer, using diplomatic skills not to be discouraging for he knows the hopes and expectations all writers share. What a shame, he thinks. Had he accepted the paper, his citation count would have increased. Now the hard work of thorough analysis lies ahead. He picks up his coffee mug and takes a large gulp.

The reviewer is a busy person, definitely time-to-result driven. A survey we regularly conduct shows that reviewers take on average 20 minutes to get a "publish or perish" impression of the paper. Naturally, Some get their first impression in a few minutes, while others read everything before forming an opinion.

And indeed, our survey also shows that reviewers only read certain parts of the paper to derive their first impression: the title, the abstract, the introduction, the conclusion, and the structure, which they discover by turning the pages and reading the headings and subheadings. They also look at some visuals and their caption. Forming a first impression on partial cursory reading is an efficient way to save time on the whole review process. Once that impression is derived, the task that follows is narrowed down to the justification of that impression:

1) If it's a good impression (**publish**) then look for all the supportive evidence that strengthens your first impression, and minimize any problem encountered by considering it minor;

2) If it's a negative impression (**perish**) then look for all the good reasons to postpone publication: identify drawbacks, inadequate data, logical errors, methodology inconsistencies, etc.

The saying is right: "First impressions count!" There is much supportive evidence for it in cognitive psychology. Enter the keywords "Halo effect" or "Confirmation bias" in your search engine, and be ready to be amazed!

In part II, the book selectively covers the parts of a paper that are read during the rapid time when the first impression is formed. This choice has been guided by the comments I received from scientists who have published papers. They stated that the methodology and results sections of their paper were the easiest and the fastest to write. It was the other parts that were difficult and took time: the abstract, the introduction, and the discussion. As to the title, the structure, and the visuals, they recognized at the end of the course that they had underestimated the key role these parts played in creating the first impression.

After reading these chapters and doing the exercises, your writing should have significantly improved. At that time, the difference between making ripples and making waves in the scientific community is less a matter of writing than it is a matter of scientific excellence, which I leave in your capable hands!

Chapter 13

Title: The Face of Your Paper

When I think about the title of a paper, quite naturally, the metaphor of a face comes to mind. So many features of a title resemble those of a face. *From your face, people get a first impression of you.* Likewise, a title contains the first words the readers will see. It gives them a first impression of how well your paper satisfies their needs and whether your paper is worth reading or not. *Your face sets expectations as to the type of person you are.* Your title also reveals what kind of paper you have written, its genre, its breadth, and its depth. *Your face is unique and memorable. It is found on your passport and various official documents.* Your title is unique. It is found in references and databases. *What makes your face unique is the way its features are harmoniously assembled.* What makes your title unique is the way its keywords are assembled to differentiate your work from the work of others.

When I was 12 years old, I stumbled upon a strange book on the shelves of my local library. It was about morphopsychology, the study of people's characters as revealed by the shape of their faces. It was fun trying to associate a face with a character. Discovering a paper from its title should also be fun. To include you in the fun, I have turned the following section into a dialogue. Imagine yourself as the scientist being asked the questions. How would you answer?

Six Titles to Learn About Titles

Author: Greetings Mr. Scientist. I'd like to introduce a series of six titles and ask you one or two questions about each one. These titles may be in areas you are not familiar with, but I'm sure you'll do fine. Are you ready?

Scientist: *By all means, go ahead!*

Author: All right then. Here is the first title.

"Gas-Assisted Powder Injection Molding (GAPIM)"[1]

Based on its title, is this paper specific or general?

Scientist: *Hmm, you are right, I know nothing about powder injection molding. The title seems halfway between being specific and being general. "Powder Injection Molding" by itself would be general, maybe a review paper. But, this title is a little more specific. It says "Gas-assisted", which tells me there are other ways to do powder injection molding.*

Author: You are right. GAPIM is used to make hollow ceramic parts. People in that field would be quite familiar with powder injection molding and its PIM acronym. What would have made the title more specific?

Scientist: *The author could have mentioned a new specific application for GAPIM.*

Author: Good, How do you feel about the use of the GAPIM acronym in the title?

Scientist: *I am not sure it is necessary. I have seen acronyms in titles before, but they were used to launch a name for a new system, a new tool, or a new database. The acronym was usually more memorable than the long name it replaced. Unless GAPIM is so well known, people have it memorized and it has become a search keyword, I don't think it should be in the title.*

Author: Thank you. How about this second title: general or specific?

"Energy-Efficient Data Gathering in Large Wireless Sensor Networks"[2]

Scientist: *This title is very specific and its scope is well-defined: it is not sensor networks, it is wireless sensor networks, more precisely large*

---

[1] Dr. Li Qingfa, Dr. Keith William, Dr. Ian Ernest Pinwill, Dr. Choy Chee Mun and Ms. Zhang Suxia, Gas-Assisted Powder Injection Moulding (GAPIM), ICMAT 2001, international conference on materials for advanced technologies, Symposium C Novel and Advanced Ceramic materials, July 2001.

[2] KeZhing Lu, LiuSheng Huang, YingYu Wan, HongLi Xu, Energy-efficient data gathering in large wireless sensor network, second International Conference on embedded software and systems (ICESS'05), Dec 2005, pp. 327–331.

wireless sensor networks. And the paper is only looking at data gathering in these networks. Its contribution, "energy efficient" is placed where it should be, right at the beginning of the title. "Energy efficient" gives me a hint that data gathering is not energy efficient in large networks. By the way, I am wondering whether "large" is the correct descriptor here, maybe "sparsely populated" would be better.

Author: You are perfectly entitled to logically infer that from the title. Actually, all readers generate hypotheses and expectations from titles. How about these two titles: are they both claiming the same thing?

"Highly efficient waveguide grating couplers using Silicon-on-Insulator"
"Silicon-on-Insulator for high output waveguide grating couplers"

Scientist: *The second title seems to introduce a new technology — Silicon-on-Insulator — to make waveguide grating couplers of the high output kind. Come to think of it, I'm not clear as to what "high output" means. It may be a type of coupler but it may also be a benefit, meaning that other technologies can only deliver lower output. You can tell I'm not an expert in this field either! Now let's look at the first title. What comes first in the title is usually the author's contribution, so this paper seems to be more concerned about making the whole system more efficient, using existing silicon-on-insulator technology. In my opinion, these are the titles of two different papers. The first paper published was the one introducing Silicon-on-insulator.*

Author: Bravo! You are doing fine. Now look at the following titles. Besides the use of an em dash or a colon to introduce the benefit of Web Services, these two titles are equivalent: which one do you prefer?

"Web services — an enabling technology for trading partners community virtual integration"[3]
"Web services: integrating virtual communities of trading partners"

Scientist: *Um… this is a difficult one. The long five word modified noun in the first title is difficult to read, yet I am attracted by the catchy term 'enabling technology' even though it isn't a search keyword. The second title is easy to read. It is shorter, clearer, more dynamic, and purposeful. But is it necessary to put a colon after "Web Services"? The second part*

---

[3] Siew Poh Lee, Han Boon Lee, Eng Wah Lee: Web Services — An Enabling Technology for Trading Partners Community Virtual Integration. ICEB 2004: pp. 727–731.

of the title does not really explain or illustrate Web services. Could the title be changed to "Integrating virtual communities of trading partners through Web Services"? In this way, what is new comes at the beginning of the title. I don't think that web services are really new.[4]

Author: The title could be changed to what you propose. You are right; the second title is more dynamic. The use of the verbal form "Integrating" makes it so. Many papers have a two part title separated by an em dash or a colon. The net effect of that separation is to create two places of emphasis in the title: before and right after the punctuation. Otherwise, only the first part of the title is emphasized. Now regarding your remark about the novelty of web services, if web services were indeed new, you could have a two part title. You are doing very well. Only two more titles. Is the following title for a British or an American journal?

*"**Vapor pressure assisted void growth and cracking of polymeric films and interfaces**"*[5]

Scientist: *Vapor with an "o". It is for an American journal, isn't it? If it had been for a British paper, they would have written "vapour". One has to be careful with keyword spelling nowadays, even if search engines are getting better. Fortunately, the title contains many keywords, so it would be easy to find.*

Author: Actually, in this field there is an alternative keyword frequently used also, but by the British. Instead of "vapor pressure-assisted", they use "moisture-induced". How would you make sure that both British and American scientists find your paper?

Scientist: *Hum,… let me think. Maybe I would use one keyword in the title, and the other one in the abstract. Since the search is often done on title and abstract, the paper has a chance to be found by both communities.*

---

[4] Our survey here again finds fascinating results. Over hundreds of sessions, about one third prefer the first title while two thirds prefer the second title. This just goes to show how damaging compound nouns are to reading clarity!

[5] L. Cheng, T.F. Guo, 2003. Vapor pressure assisted void growth and cracking of polymeric films and interfaces, *Interface Science*, 11(3): 277.

Author: That's a good idea, but it may backfire. Remember that in order for the reader to see your abstract, they would have had to click on your title, which means that they were attracted by THAT keyword, and not its synonym, which they might not know. Choosing different keywords between the title and abstract would also result in search algorithms such as google scholar giving you a lower rank! Instead, you should use the alternative keyword in the *keyword list*. That's what the list is for!

Aside from the synonymous keywords issue, is there anything else in this title you find potentially confusing?

Scientist: *This title contains two "and"s. Are there two contributions in this paper: **vapor-pressure assisted void growth** being one, and **cracking of polymeric films and interfaces** being the other, or is there only one: **vapour assists both** void growth and cracking? The second "and" is just as ambiguous: does the adjective "polymeric" apply to films and interfaces, or only to films? I am sure an expert would find the title unambiguous, but non-experts like myself lack the knowledge to disambiguate.*

Author: Excellent observation. Titles have to be clear to all, experts and non-experts. Besides *and* and *or*, other prepositions can also be quite ambiguous in titles. For example, the preposition *with* could mean *and* as in "coffee with milk", or it could mean *using* as in "stir the coffee with a spoon". Whenever you have an unclear preposition in your title, see you if you can replace it with a *through*, *for*, *using*, etc… instead.

The time has come for our last title. It is somewhat tricky. Can you identify the author's contribution?

"A new approach to blind multi-user detection based on inter symbol correlation"

Scientist: *Other researchers are already doing research in this field, and the author is following the pack with a new approach. Personally, I don't like the word 'approach' : it is vague. I would use 'method,' 'technique,' 'system,' 'algorithm,' or 'technology' instead. They are more specific. And I also don't like titles that start with 'a new' something. 'New' does not specify what is new or what makes it new. Calling the approach new can also be a lie! If the reader picks up this paper within a few years of it being published, then the author's claim of novelty is correct. But if the paper* *is being retrieved five years down the line, the reader might be surprised to find in the abstract that the method proposed is already quite dated! Regarding the contribution of this paper, I must say I am at a loss. The inter symbol correlation could be new, but if that is the case, why is it at the back of the title? It should be upfront. "Inter Symbol correlation for blind multi-user detection" is clear. It may also be — and I suspect this is the case — that inter symbol correlation is not new, but the author modified the method. That would explain the use of 'based on'. In that case, why doesn't he tell us the benefit of the modified method? Something like... "Modifying Inter Symbol Correlation to increase accuracy of blind multi-user detection"...It would be more informative and more compelling.*

Author: You are quite good at this. Thank you so much for assisting me in this dialog.

Scientist: *Not at all!*

Less time than you think

Vladimir often liked to Google his one published paper to see who cited it, or what other people wrote about it. But, that day, instead of typing the whole title, he typed only two of its main keywords, thinking it would be enough. Aghast, he looked at the first page. It was filled with other people's titles, and there were another ten pages following that one. He started scanning down the list of titles, barely spending a couple of seconds on each one. And then the thought hit him. That was what other researchers did when searching for interesting papers!

"This is terrible," he shouted.

Ruslana heard him. She asked "What is terrible, darling?"

"Good grief! I spent 9 months doing research, 2 full weeks writing my paper, but if its title does not catch the eyeballs of googling readers, my paper won't even be read, and I can kiss goodbye to citations and career growth!"

Ruslana helped as helpless people usually do, by stating the obvious.

"Well, you'd better write an attractive title then!"

Vladimir responded:

"Thanks a lot, and how am I supposed to make a title attractive, Mrs Toldoff?"

She escaped by jesting:

"Try lipstick, Darling."

Actually, Ruslana made an interesting point. The title is the face of your paper. But how do you recognize a face amongst hundreds?

- First, look at a face, face on, not from the back (see technique 1 below).
- Smile. A smiling face is more memorable than a bland one. It is engaging, alive, dynamic (see techniques 2 and 3).
- Do not hide the salient characteristic features by which people recognize your face behind scarves, glasses, or veils (see techniques 4 and 5).
- Find a way to enhance the attractiveness and uniqueness of the face through make-up, special hairstyle, mustache, etc (see technique 6).

But whatever you do, don't be an impostor and use facial cosmetic surgery to turn into another Michael or Elvis. Each face is unique, and it should remain so.

Why don't you identify how these solutions also apply to the crafting of a great title? You may even have your own solutions. Here are six techniques that I have found effective.

Six Techniques for Improving Titles

1-Placement of contribution upfront in a title

As seen in the six titles reviewed so far, readers expect to see the contribution of a paper at the same place in each title: its beginning. The contribution is followed by the scope, the method, or the application (as in the example hereafter)

Self-installing substation for insular tidal-energy farms

The majority of the titles in print are incomplete sentences lacking a conjugated verb. In certain fields however (life sciences being one), the title is a full sentence with a conjugated verb.

☀ When a title contains a conjugated verb, the contribution starts with the verb and continues until the end of the sentence (the sentence stress); otherwise, the contribution is stated upfront in the title.

"Glucocorticoid-induced thymocyte apoptosis is associated with endogenous endonuclease activation"[5]

The contribution is not that thymocyte apoptosis is induced by Glucocorticoid. This is an already known fact. The contribution is the association between apoptosis and endogenous endonuclease activation.

2-Addition of verbal forms

A title without a conjugated verb lacks energy. Nouns do not have the action-packed strength of a verb.

☀ The gerundive and infinitive verbal forms add energy to a title without a conjugated verb.

*"Data learning: **understanding** biological data"*[6]

*"Nonlinear Finite Element Simulation **to Elucidate** the Efficacy of Slit Arteriotomy for End-to-side Arterial Anastomosis in Microsurgery"*[7]

3-Adjectives and numbers featuring qualitative or quantitative aspects of the contribution

Adjectives and adverbs are also used to attract — *fast, highly efficient* or *robust*. They enhance the contribution. Since adjectives are subjective, replacing them with something more specific is always better. A "100 MHz DCT processor" is clearer than a "fast DCT processor"; and while in twenty years "fast" will make a liar out of you, "100 MHz" will not. And by the way, avoid using *'new,' 'novel,'* or *'first,'* which are not informative because a contribution should be new, novel and a first, anyway.

A 100 MHz 2-D 8 × 8 DCT/IDCT processor for HDTV applications[8]

---

[5] A. H. Wyllie Glucocorticoid-induced thymocyte apoptosis is associated with endogenous endonuclease activation, *Nature* **284**: 555–556 (10 April 1980).

[6] Brusic, V., Wilkins, J. S., Stanyon, C. A. and Zeleznikow, J. (1998a). Data learning: understanding biological data. *In:* Merrill G. and Pathak D.K. (eds.) Knowledge Sharing Across Biological and Medical Knowledge Based Systems: Papers from the 1998 AAAI Workshop pp. 12–19. AAAI Technical Report WS-98-04. AAAI Press.

[7] Reprinted from journal of biomechanics, vol 39, Hai Gu, Alvin Chua, Bien-Keem Tan, Kin Chew Hung, nonlinear finite element simulation to elucidate the efficacy of slit arteriotomy for end-to-side arterial anastomosis in microsurgery, pages 435–443, copyright 2006, with permission from Elsevier."

[8] Madisetti, A. Willson, A.N., Jr., "A 100 MHz 2-D 8 × 8 DCT/IDCT processor for HDTV applications", IEEE Transactions on Circuits and Systems for Video Technology, Apr 1995, Vol. 5, no. 2, pp. 158–165.

4-Clear and specific keywords

Specific keywords attract the expert. The specificity of a paper is proportional to the number of specific keywords in its title. Beware of keywords buried in long modified nouns whose clarity is inversely proportional to the length of the modified noun. To clarify such nouns, add prepositions. You may lose in conciseness, but you certainly gain in clarity — the gain is greater than the loss.

> *"Transient model for kinetic analysis of electric-stimulus responsive hydrogels" (unclear)*

> *"Transient model for kinetic analysis of hydrogels responsive **to** electric stimulus" (clear)*

5-Smart choice of keyword coverage

Even published, an article has little impact if not found. Readers find new articles through online keyword searches; that is why choosing effective keywords is vital.

Keywords are divided into three categories (☞1).

General keywords (*'simulation,' 'model,' 'chemical,' 'image recognition,' 'wireless network'*) are useful to describe the domain, but they have little differentiating power precisely because they frequently

Keyword type

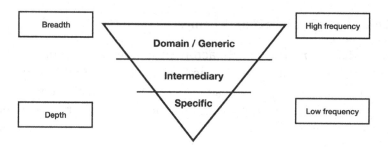

Figure ☞1
Specialized keywords are at the pointed deep end of the inverted triangle. General keywords are at the broad top end of the triangle. The general to specific scale is correlated to the frequency of use of a scientific keyword. Depth and breadth of a keyword are not intrinsic qualities. They depend on the frequency of use of these words in the journal that publishes the paper. The reader's knowledge also influences the perception of keyword levels: the less knowledgeable the reader is, the more general keywords seem specific.

appear in titles. They do not bring your title closer to the top of the list of retrieved titles. Intermediary keywords are better at differentiating. They are usually associated with methods common to several fields of research (*'fast Fourier Transform,' 'clustering,' 'microarray'*) or to large sub-domains (*'fingerprint recognition'*). But, for maximum differentiation, specific keywords are unbeatable (*'Hyper surface,' 'hop-count localization,' 'non-alternative spliced genes'*). For a given journal, or for domain experts, the category of a keyword is well-defined. It changes from journal to journal, or from experts to non-experts. *Polymer* would be an intermediary keyword in the journal Nature, but it is definitely generic in the journal of Polymer Science.

Make sure your title has keywords at more than one level of the triangle. If too specific, your title will only be found by a handful of experts in your field; it will also discourage readers with a sizable knowledge gap. If too general, your title will not be found by experts, or will only pop up on page 5 of the search results. Decision on the keyword choice is yours. Base it on who you are writing your paper for.

If you are a known author pioneering the research in your field, someone readers follow through your announcements of publications on twitter, or someone whose articles are retrieved simply by searching by author name, citations, or references... If you are that someone, do not worry excessively about other people's titles: you are the leader. But if that is not the case (yet), make sure your title has more than one keyword. People need keywords to find it.

6-Catchy attention-getting schemes

Catchy Acronym. The BLAST acronym is now a common word in bioinformatics. It started its life as five words in a title, "Basic Local Alignment Search Tool," published in the 1990 Journal of Molecular Biology. The author built a fun and memorable acronym, and everyone remembered it. Acronyms provide a shortcut to help other writers refer to your work succinctly.

> *"Visor: learning **VI**sual **S**chemas in neural networks for **O**bject **R**ecognition and scene analysis"*[9]

---

[9] Wee Kheng Leow (1994). VISOR: Learning Visual Schemas in Neural Networks for Object Recognition and Scene Analysis, PhD Dissertation; Technical Report AI-94–219, June 1994.

The title above is that of the doctoral thesis of Wee Kheng Leow. Other researchers mentioning his work could, for example, write "In the VISOR system [45]". The acronym provides a convenient way for others to refer to his work. Notice that both BLAST and VISOR are memorable. Acronyms like GLPOGN are doomed to fail.

Question. The question makes a mighty hook. But journal editors rarely allow them. If you are bold enough to propose a question in your title, the answer must definitively be found in the paper. Yes, or no. No maybes or it depends.

"Software acceleration using programmable logic: is it worth the effort?"[10]

Words out of the expected range. Here is a catchy and intriguing title that relies on a simile.

"The Diner-Waiter pattern in Distributed Control"[11]

"Distributed control" is not usually associated with the interaction between a restaurant waiter and a customer. What the title gains in interest, it loses in retrievability: it only has one general domain keyword, (*"distributed control"*) and researchers in this domain are unlikely to even think of *"diner-waiter"* as a search keyword. The next title has words out of the expected range that do not work cross-culture.

"The inflammatory macrophage: a story of Jekyll and Hyde"[12]

Would you understand this title if you were a researcher with a non-English background? Dr. Jekyll never published any papers, as far as I know. This title only makes sense if you have read the 1886 novel by Robert Louis Stevenson.[13]

---

[10] Martyn Edwards, Software acceleration using programmable logic : is it worth the effort, Proceedings of the 5th international workshop on hardware/software co-design, IEEE computer society, p. 135, March 1997.

[11] H. He and A. Aendenroomer, "Diner-Waiter Pattern in Distributed Control", Proceedings of 2nd International Conference on Industrial Informatics, (INDIN'04), Berlin, Germany, 12–16 June 2004, Vol. 2, pp. 293–297.

[12] Duffield JS., The inflammatory macrophage: a story of Jekyll and Hyde, Clin Sci (Lond). 2003 Jan; 104(1): 27–38.

[13] In the book, Dr Jekyll suffers from a personality disorder. He is a medical practitioner by day who helps and heals people, but an evil character by night who harms others and commits a murder.

Besides these six popular techniques to improve titles, listed here without example are three more.

- Some keywords become buzzwords that carry the passion of our times. Seeing them in titles attracts the reader keen to keep up-to-date with the happenings in Science. For example, some buzzwords in 2020 included Deep Learning, Additive Manufacturing, and CRISPR-Cas9.
- A shorter title is more attractive than a long one, and a general title more attractive than a specific one.
- Words that announce the unexpected, the surprising, or the refutation of something well established, all fuel the curiosity of the reader.

To make a title catchy, there is one rule only: catchy, yes; dishonest, no.

Purpose and Qualities of Titles

Purpose of the title for the reader

- It helps the reader decide whether the paper is worth reading further.
- It gives the reader a first idea of the contribution: a new method, chemical, reaction, application, preparation, compound, mechanism, process, algorithm, or system.
- It provides clues on the paper's purpose (a review, an introductory paper), its specificity (narrow or broad), its theoretical level, and its nature (simulation, experimental). By the same means, it helps the reader assess the knowledge depth required to benefit from the paper.
- It informs on the scope of the research, and possibly on the impact of the contribution.

Purpose of the title for the writer

- It allows the writer to place enough keywords for search engines to find the title.
- It catches the attention of the reader.
- It states the contribution in a concise manner.
- It differentiates the title from other titles.
- It attracts the targeted readers, and filters out the un-targeted readers.

Qualities of a title

UNIQUE. It differentiates your title from all others.

LASTING. Do not to use *new* in a title that outlives you.

CONCISE. Remove redundant or filler words ('study of'). If your title is already unique and easy to find without the details, remove them.

CLEAR. Avoid long modified nouns. They bring imprecision and misunderstanding.

HONEST and REPRESENTATIVE of the contribution. It sets expectations and it fulfills them.

CATCHY. The writer has one chance and 2 seconds to interest the reader.

EASY TO FIND. Its keywords are carefully chosen.

What do you think of your title? Does it have enough of the qualities mentioned here? Is your contribution featured at the head of your title? It is time you take a closer look.

Title Q&A

Q: When do I write my title?

A: There is no set time. You may write a tentative title as soon as you identify something is novel and useful enough to merit publication. An early title gives focus to your work. As your research progresses, you may decide to change your title to include new findings. But you may also discover new elements that, on their own, represent sufficient contribution to merit their own paper. How would you know which route to take? Two papers may be better if your title gets longer when you try to bring in everything together. But if your title becomes shorter, and more general, you may have a better, more attractive paper. General or generic titles should not be abused, however. It is better to have a string of published short papers than to have an unpublished paper with a generic title covering a shapeless amalgam of new research tidbits.

Q: Should I look at other titles in my list of references before or after I write my final title?

A: Your final title should have all the qualities mentioned in this chapter, and not be influenced by other people's titles — apart from making sure your title is unique. This said, while looking at the titles of related papers, you may discover that the most highly cited papers use a particular keyword, whereas you use a synonymous keyword also often found. It may be advantageous to use instead the one found in the title of the highly cited papers so that when their titles are retrieved, your title is retrieved alongside theirs.

Q: When two keywords are synonymous, which one do I choose for my title?

A: From what we have seen before, the most frequent one should be chosen. When two different keywords with the same meaning appear with the same frequency in titles, choose one for the title and abstract and put the other in the keyword list. That way, search engines will find your paper, regardless of the keyword used for the search.

Q: Is it appropriate to mention the impact of the contribution in a title?

A: *"Multiple reflections of Lamb waves at a delamination"* is the title of a paper by T. Hayashi and Koichiro Kawashima.[14] After reading the paper, scientists who attended the writing skills seminar suggested to rewrite the title to depict more precisely what the paper contains. Here is their proposed title:

Multiple reflections of Lamb waves for rapid measurement of delamination.

---

[14] T Hayashi, Koichiro Kawashima, 2002. Multiple reflections of Lamb waves at a delimitation, *Ultrasonics*, **40**(1-8): 193–197.

Note the addition of *'rapid measurement'*. What attracted the readers, the research impact, was absent from the title. Yes, the title is longer, but also more attractive.

Q: Isn't it better to keep the title general so that it attracts a maximum of readers?

A: The following title of a paper by B. Seifert *et al.*, published in the artificial organs journal (Volume 26 Issue 2, Pages 189–199) is a typical title with two parts separated by punctuation. It emphasizes the name of the polymer and its novelty in membrane-forming.

"Polyetherimide: A New Membrane-Forming Polymer for Biomedical Applications"

The terms *'biomedical applications'* is general, maybe too much so for a journal dealing with biomedical applications anyway. The indefinite article *'a,'* is general and non-descriptive: it means *'one of many'*. *'New'* is going to be obsolete very soon. *'Membrane'* is also a general term. Membranes are of many types and have many properties. Here is how the readers of this paper proposed to change its title to make it more specific and yet more attractive:

Polyetherimide: A biocompatible polymer to form anti-fouling membranes

They added a hot keyword *'Biocompatible,'* and they characterized the membrane property with *'anti-fouling'*. The title now includes a dynamic verb (*'to form'*) enhancing its appeal.

Q: I agree that a short title is better than a long one, but how does one go from a long title to a short title?

A: The following title of a paper by Cook BL, Ernberg KE, Chung H, Zhang S, published in the PLoS One. 2008 Aug 6;3(8) is long:

"Study of a synthetic human olfactory receptor 17–4: expression and purification from an inducible mammalian cell line"

Look out for words such as *'study of,'* or *'investigation of'*. You can safely remove them. *'a'* is not attractive, and misleading because in this case, there is only one such receptor. Readers of the article rewrote the title to make it concise by removing what they considered an unimportant and unproductive search keyword (the Mammalian cell line).

Purifying the synthetic human olfactory receptor 17–4 in milligram quantities

The readers placed the outstanding contribution right upfront in the title, in verb form to make the title more attractive, and they added the outcome of the contribution at the end of the title.

Q: What are the consequences of changing the keywords in my title?

A: Here are two alternative titles for a same paper written by Dr. Linda Y.L. Wu, A.M. Soutar and X.T. Zeng, and published in Surface and Coatings Technology, 198(1–3), pp. 420–424 (2005).

"Increasing hydrophobicity of sol-gel hard coatings by chemical and morphological modifications"

"Increasing hydrophobicity of sol-gel hard coatings by mimicking the lotus leaf morphology"

The second title is quite catchy. Only one keyword describing the methodology was lost (*'chemical modification'*). *'Lotus leaf'* is unexpected. That keyword may attract scientists outside the domain of manufacturing technology, and even journalists writing for widely distributed science magazines. The author of this paper, Dr. Linda Wu, decided to keep the first title even though the second title was indeed attractive. Why? The journal she targeted with her title was "Surface and Coating Technologies". Its readers have a chemical and material science background, not biology. Had she chosen the *'lotus leaf'* title, she would have attracted a more general audience to her paper. As a result, she would have had to rewrite the whole paper, change its structure (emphasis on morphology and biomimicry), and simplify the vocabulary (simpler terms understandable by a larger group of non-experts).

Q: If I submit a paper to different journals, do I need to change the title?

A: First, if you write a paper, it is for *one* journal. Look at the web page of the journal you are targeting. It specifically mentions the type of contents the journal covers. Each journal attracts different types of readers in terms of expertise. Therefore, each paper is written for the specific readers of a specific journal. A paper cannot possibly be written the same way for two different journals. Not only should your title change, but also the contents of the paper.

Also, one should not send the same paper to two journals at the same time. Submission should be a serial process. Pick one journal. If that journal decides not to publish you or if you feel that the journal's lengthy review process (due in part to a reviewer's attitude towards your paper) delays important results from being published, submit it to another journal — after reworking its contents and title to match the readers' interests.

Q: In the role of the title for the writer, you write "[the title] catches the attention of the reader **targeted** by the writer". Why should the writer target anyone? Isn't the reader the one identifying the target according to his or her needs?

A: Your impact will be negligible if you do not have in mind the reader who can make use of your contribution. What keywords is this reader looking for? Are they in your title? These keywords usually describe the impact or the application domain. In a precedent title, the words *'in milligram quantities'* were added to the title by the readers themselves, as they thought this was very significant and opened many new applications. In another title the words *'rapid measurement'* were added by readers, again reflecting the same concern to present real solutions to real problems faced by the scientists in this field. Put yourself in the shoes of your reader, and if space allows, find a way to mention the outcome of your contribution in the title of your paper.

Q: In a previous chapter you mentioned that adjectives make claims and are therefore dangerous to use in a scientific paper. But in this chapter you write that adjectives are attractive. Isn't there a contradiction here?

A: Precisely, the eye-catching adjectives make claims. But so does any title. It should not be astonishing to find an adjective such as *'robust,'* *'efficient,'* or *'fast'* in a title. Once the claim is made, the writer is expected to provide early justification for that claim in the abstract, even if the justification is only partial.

Q: When one writes a title, what are the advantages and drawbacks for being first in a field?

A: The advantage is clear. If you are a pioneer in your field, the choice of words is entirely yours. Since you are the first to write in this field, you need not worry about titles that may already have been used. Think about it. Imagine being the first to write about dialogue in speech recognition. Finding a title is easy: *"Dialogue in speech recognition"*. Nice and short. Now imagine you are writer 856 with a paper in this crowded field. You have to be much more specific to differentiate your title from all the others. As a result, you might have to settle for a long specific title like *"semantic-based model for multi phase parsing of spontaneous speech in dialogue systems"*

Being first gives you the opportunity to have a shorter title... that won't be found! You are a pathfinder. At first, people may not find you through the possibly new keyword you introduce (the name you gave to a new polymer, for example), because they are not yet aware of it. This is why you want to make sure your title is found by including other well-known keywords. Alternatively, if you are already famous, you need not care. People find you by name, not through the keywords of your papers.

Q: What evidence supports your claim that you should place what is novel upfront in the title?

A: Eye tracking studies, studying how people search through lists, have identified that people spend more time reading the beginning of each item in a list than the end of that item. Often, when the beginning does not capture the attention, the rest of the list item is simply skipped and the reader moves to the next line. So if you put what is interesting at the end of your long title, the reader may not even get to it.

Title Metrics

✓(+) The words representative of your contribution are upfront.

✓(+) The title has 2+ search keywords, covering different sections on the inverted pyramid.

✓(+) Your title has attractive words (non search keywords).

✓(+) No noun phrase exceeds three words.

✓(+) All your title search keywords are found in your abstract.

✓(+) Your title is read in less than two seconds and is clear at first reading.

✓(+) No search keyword present only in the abstract appears with a frequency higher than any title search keyword found in the abstract.

✓(+) The impact or outcome of your contribution is identifiable in the title.

✓(+) Your title clearly sets the scope of the research.

✓(–) The words representative of your contribution are back and front, or at the back.

✓(–) All your keywords are in the same inverted pyramid section.

✓(–) The title has only one search keyword.

✓(–) Your title has ambiguous prepositions (and, with).

✓(–) Your title has no attractive words.

✓(–) Your title contains 'a,' 'an,' 'study,' 'Investigation'.

✓(–) Nouns phrases have 3+ words (e.g. Metastable transition metal nitride coatings).

✓(–) Some title search keywords are missing from your abstract.

✓(–) Your title requires more than two seconds to read.

✓(–) Search keywords repeatedly found in the abstract are not in the title.

✓(–) Your title creates diverging expectations about its contents.

✓(–) Your title does not set the scope of the research, or does so partially.

AND NOW FOR THE BONUS POINTS:

✓(+++) The title creates only one expectation about its contents, and fulfills it.

Chapter 14

Abstract: The Heart of Your Paper

The heart plays an essential role in the human body. Similarly, the essence of an article is its abstract. *The heart has four chambers.* The abstract is also composed of four easily identifiable parts.

> ### Visuals in Abstracts?
>
> *Never say never! I used to think that abstracts had no visuals, but it looks as though I was mistaken. The table of contents of some journals (e.g. Advanced Materials, Journal of the American Chemical Society) now include one key visual alongside an abridged abstract. Is this a preview of the shape of things to come for all journals? I believe it is. A good figure far exceeds plain text in illustrating and explaining a contribution efficiently and concisely. Therefore, take note and prepare yourself. Which visual will be 'the one' to choose for your abstract?*

The abstract dissected here is at the crossroads between surgery and computer science. It comes from a paper on slit arteriotomy. The easiest way to explain arteriotomy is to visualize the surgical connection of two tubes (here, arteries). Normally the surgeon cuts an elliptic hole (with removal of material) in the recipient artery and then stitches the donor artery over the hole. In this case, however, only a slit is cut in the side of the recipient artery before the donor artery is stitched over it. No material is removed.

Does slit arteriotomy work as well as hole arteriotomy? Even if the answer is yes, surgeons are conservative. If an established procedure (the hole arteriotomy) works, why is there a need to replace it with a new one (slit arteriotomy)! Initial statistics that establish the equivalence of both techniques are not enough. What will happen to the slit in ten years when arteries age or when the patient's blood pressure rises? To find out the safety and efficacy of the new technique over time, the inventor surgeon asked for the help of computer-modeling scientists. The technique was modeled, and a paper[1] was published in the *Journal of Biomechanics*. The readers of this journal come from diverse horizons: life science, engineering science, and computer science. When they glanced at the table of contents of volume 39 of the journal, they saw the following title:

Nonlinear Finite Element Simulation to Elucidate the Efficacy of Slit Arteriotomy for End-to-side Arterial Anastomosis in Microsurgery[2]

The title has two parts: contribution and background. If you were to insert a dividing bar | between these two parts, where would you place it? The answer will come later, after you read the abstract. Note that the words in bold are common to both the abstract and the title.

"[Part 0–20 words] **The slit arteriotomy for end-to-side arterial Microanastomosis** *is a technique used to revascularize free flaps in reconstructive surgery. [Part 1–41 words] Does a slit open to a width sufficient for blood supply? How is the slit opening affected by factors such as arterial wall thickness and material stiffness? To answer these questions we propose* **a non-linear finite element** *procedure to simulate the operation. [Part 2–10 words] Through modeling the arteries using hyperelastic shell elements, our* **simulation** *[Part 3–112 words] reveals that the slit opens to a width even larger than the original diameter of the donor artery, allowing sufficient blood supply. It also identifies two factors that explain the opening of the slit: blood pressure which is predominant in most cases, and the forces applied to the slit by the donor artery. During simulation, when we increase the donor artery thickness and stiffness, it is found that the contribution of blood pressure to the slit opening decreases while that of the forces applied by the donor artery increases. This result indicates that sometimes the forces by the donor artery can play an even more significant role than the blood pressure factor.*

---

[1] Reprinted from journal of biomechanics, Vol 39, Hai Gu, Alvin Chua, Bien-Keem Tan, Kin Chew Hung, nonlinear finite element simulation to elucidate the efficacy of slit arteriotomy for end-to-side arterial anastomosis in microsurgery, pages 435–443, copyright 2006, with permission from Elsevier.
[2] H. Gu[a], A.W.C Chua[b], B. K. Tan[b], K.C. Hung[a].

*[Part 4–28 words] Our simulation **elucidates the efficacy** of the slit arteriotomy. It improves our understanding of the interplay between blood pressure and donor vessel factors in keeping the slit open. [Total: 211 words].*

Where does the bar "|" fall, in other words, where is the separation between the contribution and the context of the contribution?

"Nonlinear Finite Element Simulation to Elucidate the Efficacy of Slit Arteriotomy | for End-to-side Arterial Anastomosis in Microsurgery"

In our sample abstract, if one locates the contribution based on the word count in each part, it seems that part 3, the elucidation of the efficacy, covers the contribution (112 words). Part 2, the non-linear finite element analysis, plays an incidental role (only 10 words), yet it comes right upfront in the title. The title could have been the following:

Elucidating the Efficacy of Slit Arteriotomy | for End-to-side Arterial Anastomosis in Microsurgery using non linear finite element simulation

However, after examining the structure of the paper (headings and subheadings), it appears that the contribution is indeed the nonlinear finite element simulation. The title tallies with the structure, less so with the abstract. One concludes that the abstract is aimed at surgeons who care little about the technical details of the finite element simulation. They may never read the paper, and be content reading the abstract. Had the paper targeted computer scientists, the methodology part would have been longer and the results part shorter.

The Four Parts

Each of the four parts in the abstract (separated by word count) answers key reader questions.

Part 1: What is the problem? What is the topic, the aim of this paper?

Part 2: How is the problem solved, the aim achieved (methodology)?

Part 3: What are the specific results? How well is the problem solved?

Part 4: So what? How useful is this to Science and to the reader?

You may have noticed that our sample abstract has a **Part 0**. It is optional, and not recommended, apart from situations such as the one arising in our sample abstract. The writer, anticipating that the meaning of a keyword in the title may be obscure to non-surgeons, provides just-in-time background — in this case, a functional definition of the surgical procedure.

A four-part abstract should be the norm, but many have only three parts: the fourth one (the impact) is missing. Why?

1) It could be that the author reached the maximum number of words too early. Some authors ramble on about the need for a solution in their abstract, but then run out of space describing the benefits of that solution.

2) Did the author (mistakenly) consider that the results speak for themselves? Again, it is worth repeating that the writer must have a reader in mind when writing a paper. Who is likely to benefit from the research? The impact of the paper is what convinces a reader to download your paper. And that reader may not have enough knowledge to determine the impact (outcomes) made possible by the results (outputs).

3) Could it be that the author was unable to assess the impact as a result of the myopia caused by the atomization of research tasks among many researchers?

4) Could it be that the author was unable to mention any impact because the contribution is only a small improvement over a previous result, not enough to claim significant impact?

Whatever the reason, having less than four parts reduces the informative value of the abstract and, therefore, its value to the reader. Since the reader decides whether to read the rest of your article or not based on the abstract, its incompleteness reduces your chances to be read... and cited.

 Read your abstract and locate its various parts. Does your abstract have its four essential parts? Are the parts with the largest number of words, those corresponding to the contribution? Are you still using adjectives and remaining vague when you should be precise?

Coherence Between Abstract and Title

A rapid keyword count will determine whether the abstract is coherent with the title. The title contains 9 search keywords: [Slit, Arteriotomy, End-to-side, Arterial, Anastomosis, Microsurgery, non-linear, finite-element, simulation]. In this count, articles (a, an, the, etc.), prepositions (to, of, for, in), and non-search keywords (elucidate, efficacy) are not taken into account.

In our sample abstract, 6 words are both in the title and in the first sentence of the abstract (66%).

> (TITLE) *Nonlinear Finite Element Simulation to Elucidate the Efficacy of **Slit Arteriotomy** for **End-to-side Arterial Anastomosis** in Microsurgery* (title)
>
> (FIRST SENTENCE) *The **slit arteriotomy** for **end-to-side arterial** Microanastomosis is a technique used to revascularize free flaps in reconstructive **surgery***.

This percentage is good. Why? The reader, having just read the title, expects to know more about it as soon as possible. Can you imagine **the first sentence** of your abstract disconnected from the message announced by the title? It is unimaginable. Coherence between title and abstract is achieved through the repetition of keywords. Percentages outside the 30%–80% range should be examined more closely.

0%–20% There is a problem. The first sentence deals with generalities loosely related to the topic of the paper. It contains two title words or less. It sets the background to the problem. If it briefly explains one or two unusual title keywords, this is fine, as long as sentences 2 and 3 mention most of the other title words. Otherwise, the background is too long and, as a result, the abstract lacks conciseness.

80%–100%. Idyllic percentage? Not necessarily. The first sentence repeats the title with just a verb added. Why repeat! The first sentence should expand the title. However, if that sentence contains many more words than the title, then 80–100% may be acceptable.

To summarize, your title merely whets the appetite of your readers; they expect to know more about your title in your abstract. You should satisfy their expectation and rapidly provide more precise details. The first sentence of your abstract should contain at least one-fourth of the words in your title.

First count the total number of significant (search) words in your title (in your count, do not include non search words such as *on, the, a, in,* or adjectives). Let's call that number T.

Then, see if your first sentence contains any of the T words. If you find some, underline them IN THE TITLE. Modified forms (a noun changed to a verb or vice versa) are acceptable but synonyms are not. For example, simulation would be considered the same as simulated, but abrasion would not be the same as corrosion.

Count the number of words underlined in your title. Let's call that number U.

Calculate the percentage 100 × U/T.

What is your percentage? Between 20% and 80%, you are doing fine. Outside this range, investigate.

A second count will help you identify the strength of the cohesion between abstract and title. Are ALL title search keywords also in the abstract? They should be.

In our sample abstract, 6 out of the 9 words were in the first sentence, three were still missing. But they were in the rest of the abstract (*we propose **a non-linear finite-element** procedure to **simulate** the operation*). All title search keywords were in the abstract.

Think about it. You give high visibility to a word by giving it title status — the highest status in a paper. Why would title words be missing in the abstract? It may be for the following reasons:

- The title word is not important. Remove it from the title to increase conciseness.
- The title word missing in your abstract is really important. Find a place for it in your abstract.
- It may also be that your abstract frequently refers to a keyword not found in the title. Rewrite your title to incorporate that keyword.
- You used a synonym to avoid repetition. Don't. Repeating a title word in the abstract increases the relevance score calculated by search engines for that keyword. As a result, your title will be brought up towards the top of the list of titles retrieved.

You have already counted, T, the number of significant words in your title. Read your abstract and see if any of the important title words are missing. If some are, ask yourself why. Are your title claims too broad? Is your title not concise enough? Are you using synonyms that dilute the strength of your keywords and confuse the reader? Decide which reason applies, and modify title or abstract if necessary.

You now have three techniques to gauge the quality of your abstract.

A) Abstracts have four parts (and one optional part for just-in-time explanation of obscure title words as seen in our sample abstract). The part that represents your contribution should be the most developed.
B) Abstracts repeat all title search keywords.
C) Abstracts expand the title in the first one or two sentences because the reader expects it.

Tense of Verbs and Precision

So far, you have managed to bring the reader past the title barrier and into your abstract. Congratulations! But if the reader stops there, what are your chances to increase your citation count? You need to use your abstract as a launch pad from which the reader will propel himself or herself inside your paper by clicking on the *download pdf file* button. For some, it even means entering their visa card details for the online payment if their library does not subscribe to the journal which publishes your paper.

How does one turn an abstract into a launch pad? In two ways: by writing it in the tense used in ads — the present tense; and by convincing the reader to read more of your paper through a brief but precise account of your main accomplishments.

Accomplishments are better communicated with the dynamic present tense than with the dull past tense. There are advantages to choosing the present tense for the abstract. The present tense is vibrant, lively, engaging, leading, contemporary, and fresh. The present tense is the tense of facts: the results you demonstrate in your paper are just as true in the future your reader inhabits as the present at which you write! Some authors struggle to write in the present tense because they argue that the past tense is a more accurate representation of the research conducted. And indeed, it is true that the experiment *was done*, but the results of that experiment cannot be said to *have been true*, they ARE true, and will remain so. Substituting the past tense for the present tense can be as simple as replacing *in this study, we found the binding force between x and y to be 3.8ev* with *in this study, we FIND the binding force between x and y to be 3.8ev*. Even if the reader discovers your paper ten years after it has been published, it will feel as fresh and current as if it had just been written. The past tense is passé, déjà vu, a thing of the past, gone, stale, unexciting, and lagging. It feels like reading old news. It could even introduce ambiguity. For example, the phrase *was studied* creates a doubt: did the writer publish this before?

If ever you needed one more reason to be convinced that the present tense is preferable, first check whether the journal targeted by your paper is agreeable to such modern practices or not. Following which, if there is no journal-imposed past tense directive, think how often you jump straight from the abstract to the conclusions when you read a new paper. Since the conclusions are also written in the past tense, to the reader, reading the conclusions feels like reading the abstract all over again (i.e. boring).

Let's now move to the second element of our launch pad: precision. The title has very few words, so it needs to attract the reader with eye-catching adjectives like *robust, effective, rapid*; but such adjectives have a limit to their efficacy. If they are not backed up with precision and details, the interested reader will quickly look away. For example, which one of the two following sentences would convince you to buy the car I'm selling? *This car goes very fast. How fast? Very, very fast. Incredibly fast, even.* OR *This car goes very fast. How fast? From 0 to 60 km/h in less than six seconds, with a max speed of 370 km/h.* Precision is convincing.

The following title contains a couple of attractive words: '*clinically distinct*'. How *distinct*, though? The reader wants to know with precision before deciding whether your paper is worth reading further.

> "*A gene expression-based method to diagnose **clinically distinct** subgroups of diffuse large B Cell Lymphoma*"

The abstract does answer the question by justifying the claim made in the title.

*"**The GCB and ABC DLBCL subgroups** identified in this data set **have significantly different 5-yr survival rates after the multiagent chemotherapy (62% vs. 26%; P = 0.0051)**, in accord with analyses of other DLBCL cohorts. These results demonstrate the ability of this gene expression-based predictor to classify DLBCLs into biologically and clinically distinct subgroups irrespective of the method used to measure gene expression."*[3]

The sample abstract provided in this chapter is entirely written in the present tense. Its results are not numerical, but the main results are described with precision.

"The slit opens to a width even larger than the original diameter of the donor artery, allowing sufficient blood supply. It also identifies two factors that explain the opening of the slit: blood pressure which is predominant in most cases, and the forces applied to the slit by the donor artery. [...] the contribution of blood pressure to the slit opening decreases while that of the forces applied by the donor artery increases."

US$ 35

Vladimir had located what looked like a really interesting paper. He had asked the librarian to download it. Instead of receiving in the internal mail the paper he expected, he received an email from the librarian informing him that the research center did not subscribe to the journal that published his requested paper. However, she offered to download the paper and charge his department the download fee of US$ 35 as long as he secured his manager's approval.

"Thirty-five bucks!" Vladimir exclaimed.

His friend in the next cubicle heard him across the partition. He shouted: "Not enough, but I'll settle for fifty!"

"Oh, shut up John. The librarian is charging us thirty-five bucks for a reprint of someone's paper. I could buy a book for that price!"

John had stood up and was now looking at Vladimir across the cubicle divider.

"You must be joking! That much?" He said.

---

[3] George Wright, Bruce Tan, Andreas Rosenwald, Elain Hurt, Adrian Wiestner, and Louis M.Staudt, a gene expression-based method to diagnose clinically distinct subgroups of diffuse large B Cell Lymphoma, Proceedings of the national Academy of Sciences, August 19, 2003 Vol 100 No.17, www.pnas.org, copyright 2003 National Academy of Sciences, U.S.A.

"How much?" That was Cheng Jia, another colleague who was sitting back to back with John in the same cubicle. She appeared next to John.

The commotion attracted the attention of Professor Popov who, at that instant, happened to look away from his desk, past his opened office door, towards the group. He joined in the conversation, listened to the complaint, and advised:

"Vlad, read that abstract again. If it contains enough detail to convince you that you could really benefit from reading this paper, then I will authorize the purchase."

Purpose and Qualities of Abstracts

Purpose of the abstract for the reader

- It clarifies the title.
- It provides convincing details on the writer's scientific contribution.
- It helps the reader decide whether the rest of the article is worth buying or downloading for further reading.
- It helps the reader rapidly gather competitive intelligence.

Purpose of the abstract for the writer

- Because it has more keywords than the title, the abstract allows the paper to be found more easily.
- It states the writer's contribution in more precise detail than the title to persuade the reader to read the rest of the paper.
- When written early in the life of a paper, it guides the writing and keeps the writer focused.

The abstract is NOT to be used for the following:

1) To mention the work of other researchers, except when your paper is an extension of a (single) previously published paper, yours or that of another author.
2) To justify why the problem you have chosen is significant: the significance of your impact on the problem is what really matters.

Qualities of an abstract

TIED TO TITLE. All title search keywords are in the abstract.

COMPLETE. It has four parts (what, how, results, impact) with an optional (short) background.

CONCISE. Not longer than necessary, as a courtesy to the reader. Research is justified through significance of results, not significance of problem.

STAND-ALONE. It lives by itself in its own world: databases of abstracts, journal abstracts. It may include the main key visual in the future.

REPRESENTATIVE of the contribution of the paper. It sets expectations. It encourages the reader to read the paper.

PRECISE when describing contribution.

PRESENT. Real. News.

What do you think of your abstract? Does it have enough of the qualities mentioned here? Is the contribution you mention in your abstract consistent with that claimed by the title? A quality abstract makes a good first impression. Spend some time reviewing it.

Abstract Q&A

Q: What is an extended abstract?

A: An extended abstract is not an abstract. It is a short paper that a reader reads in less than an hour — usually a fifth of a regular paper. Unlike the abstract, it contains a few key visuals (space allowing) and references. It is short on details. For example, even though the extended abstract has an introduction, it does not paint the context of the contribution in great detail, but focuses instead on the immediate related works that are essential to understand and value your contribution. Likewise, its conclusion (if there is one) skips the future work. In some ways, writing an extended abstract is more difficult than writing a full paper because one has to make hard choices when it comes to details: what to leave out, and what to keep in to interest and convince.

Q: Can I submit to a journal a paper that corresponds to the extended abstract I prepared for a conference?

A: The short answer is no. There would be no added novelty, just additional details on the same contribution. This is an issue that journals deal with explicitly in their guides to authors, so do read your journal guidelines on this matter. When you submit your paper to non open-access journals, you sign a legally binding submission declaration assigning the copyright of your material to the journal. Therefore, reusing in the journal paper paragraphs recycled from the extended abstract would be considered self-plagiarism.

The extended abstract is often nothing more than the paper corresponding to the first research step. Additional steps are required for a journal paper. Its title would be different, its visuals also, and much more.

It is always good to declare the existence of your extended abstract to the journal editor when submitting your paper, and to explain the differences between it and your paper (different methodology, increased scope, new findings...).

Q: Do all abstracts have four parts?

A: Not all abstracts have four parts, sometimes with good reason. A review paper that covers the state of the art in a particular domain has only one or two parts. Short papers (letters, reports) have one or two lines. "Extended" abstracts are written prior to a conference, in some cases well before the research is completed; as a result, their parts 3 and 4 are shallow or missing. But, apart from these special cases, all abstracts should have four parts.

Q: Some journals limit the length of an abstract to 300 words. Is it best to use all 300 words?

A: Think of the busy reader! If you can write your abstract in less than 300 words, and still have a complete abstract (4 parts), then be brief! Brevity comes through selection.

Q: Abstracts should be written in the past tense. Everybody does it this way. Where do you get this present-tense nonsense from?

A: Never say never! I used to think that abstracts were only written in the past tense because they refer to work that is completed. It looks as though I was mistaken. More journals now accept and even recommend the use of the present tense for the abstract. Even NASA scientists are advised to use the present tense in their abstracts.

Before you select the tense for your abstract, read the journal recommendations to the author. If the journal does not forbid the use of the present tense, consider using it.

Q: Does an abstract contain text only?

A: Actually, the visual abstract is slowly making its way into some chemistry, material science, or even biology journals. It helps the reader understand the contribution visually and with few words. Pathways, chemical formulas, or microstructures are more effectively understood visually. If your journal demands a visual for its table of contents, be ready to pick the visual the most representative of your contribution.

Q: My paper is a review. Do the principles detailed in this chapter also apply to this kind of paper?

A: A review is quite different. Look at the resources provided in the final chapter to find guidelines for review papers.

Q: Should I mention the method if it is a very common one used in my field?

A: Yes, but in that case, do not elaborate, half a sentence will do. If you modified that method, just indicate how it differs from the standard one.

Abstract Metrics

✓(+) All search keywords in your title are also in the abstract.

✓(+) Your abstract has all four main parts (what, how, results, impact).

✓(+) The part that contains the contribution has the greatest word count.

✓(+) Your abstract does not contain background or justification of problem importance.

✓(+) Your abstract is written using verbs at the present or present perfect tense only.

✓(+) Your abstract mentions the main result(s) with precision or the key method steps if the contribution is a method.

✓(+) By revealing the main outcome of your results, your abstract targets the reader who stands to benefit the most from your research.

✓(–) One or more search keywords in your title are missing in the abstract.

✓(–) The first sentence in your abstract is more or less a repetition of the title.

✓(–) The first sentence in your abstract contains none or just one of the title keywords.

✓(–) Your abstract is missing one of the main parts.

✓(–) The part that contains the contribution does not have the greatest word count.

✓(–) Your abstract is written using the past tense only, or a mix of various tenses.

✓(–) Your abstract remains vague and lacks precision when mentioning the main result(s) or the key method steps if contribution is a method.

✓(–) The abstract remains vague on part 4 (the impact) because no reader is targeted.

AND NOW FOR THE BONUS POINTS:

✓(+++) The reader is able to figure out the title of your paper, just by reading your abstract.

Chapter 15

Headings–Subheadings: The Skeleton of Your Paper

The skeleton gives a frame to the body. With it, the reinforced body takes shape; without it, the human would be a jellyfish. The skeleton of a paper is its structure. *The skeleton supports the various parts of the body according to their functional needs.* Composed of headings and subheadings set in a logical order, the structure reinforces the scientific contribution. *The skeleton is standard but it allows for variations in shape and size.* Headings are generally the same from one article to the next (introduction, discussion, conclusions) but subheadings differ. *The most sophisticated parts of the skeleton are also the most detailed (backbone, metacarpus, metatarsus).* The most detailed part of a structure contains the largest amount of contributive details.

> ### The Scientific Paper: 300 years of History
>
> *In an article published in The Scientist entitled "What's right about scientific writing," authors Alan Gross and Joseph Harmon defend the structure of the scientific paper against those who claim it does not represent the way science "happens". The structure, refined over more than 300 years, has enabled readers to evaluate the trustworthiness and importance of the presented facts and conclusions. The authors praise the standard narrative. They also observe that today, as a result of the increased role played by visuals, it is necessary to go beyond the interpretation of linear text.*

Structures for Readers and Structures for Writers

Before they begin writing, many writers think deeply on their content to plan out what they're going to say, and where they're going to say it. They divide their content into little sections, each of which is separated by a heading or subheading. Subsequently, they use these headings to remind themselves of the general topic that the content should follow. This system makes for organised writing, and this is a good thing. But is it enough?

As usual, this type of writing is writer-focused. The structure is created to help the writer organise and develop their thoughts. But if there is one thing you should be hyper-aware of by this point, it is that the writer needs to be first and foremost reader-focused. To understand how to best write a structure for the reader, you must first understand how readers use structures.

So ask yourself the question: as a reader, how do you use the structure? Do you carefully read through all of it, memorising each line, trying to reconstruct the mental model of the writer's logic for the paper? Or do you instead browse over the headings and subheadings until a keyword catches your eye, prompting you to go straight to that part of the paper? You see, readers use the structure not as a skeleton, but as a collection of potential entry points. Readers may identify two headings of interest from the structure and read only the paragraphs under these headings before moving on to something else! A good structure allows them to skip through your paper effectively, instead of having to read large chunks of text that may be irrelevant to their needs or interests. A good structure is a writer's nod of understanding to the time-pressed reader.

Four Principles for a Good Structure

A structure that plays its role follows these four principles:

1. Contribution guides its shape.
2. Headings and subheadings detailing the contribution are grouped.
3. Title words reflective of the contribution (not necessarily all title words) are repeated in its headings and subheadings.
4. It tells a story clear and complete in its broad lines.

Studying the structure of your paper allows you to identify important problems, such as an imperfect title, or a paper which is too complex, too detailed, too premature or too shallow.

Let us review the structure of the paper on slit arteriotomy. The words in italic are common to both title and structure.

Non-linear Finite Element Simulation to Elucidate the Efficacy of Slit Arteriotomy for End-to-side Arterial Anastomosis in Microsurgery[1]

1. Introduction
2. Mechanical factors underlying *slit* opening
3. Methodology for computer *simulation*
   3.1 Reference configuration for the *finite element* model
   3.2 Geometry details and boundary conditions of *the finite element* model in the reference configuration
   3.3 Hyperelastic material for the *arteries*
   3.4 *Simulation* procedure for the operation
4. Results and discussion
5. Conclusions

Recall that a title has two parts: the front part represents its contribution; and the back part, its context.

"Non-linear Finite Element Simulation to Elucidate the Efficacy of Slit Arteriotomy [Contribution] for End-to-side Arterial Anastomosis in Microsurgery [Context]"

Principle 1: the contribution guides the shape of a structure

In the example above, three headings are standard: *Introduction, Results & discussion,* and *Conclusions.* Standard headings are disconnected from titles; they contain no title word. They simply mark the location and the function of a part. In contrast, headings 2 and 3 are meaningful: they contain nearly half the title words, and relate directly to the contribution.

---

[1] Reprinted from journal of biomechanics, Vol 39, Hai GU, Alvin Chua, Bien-Keem Tan, Kin Chew Hung, nonlinear finite element simulation to elucidate the efficacy of slit arteriotomy for end-to-side arterial anastomosis in microsurgery, pages 435–443, copyright 2006, with permission from Elsevier."

Heading 3 dominates this structure. With four subheadings, it provides much detail on the contribution. The subheadings organize the details in a logical order. All of this is to be expected, is it not? A structure should be the most detailed where the author has the most to write about, namely the scientific contribution of the paper. The structure has to expand to match the level of detail by offering more subheadings to help organize these details in a logical order, for the benefit of the reader and for the sake of clarity (☞1).

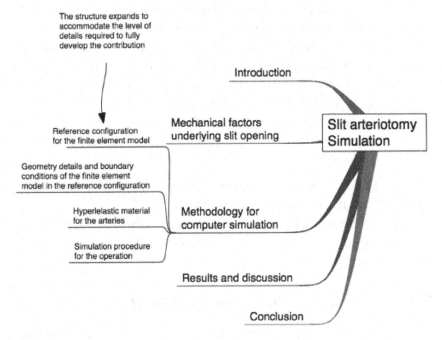

Figure ☞1
Contribution is often found under the heading which has the deepest level of indentation, and the largest number of subheadings.

This first principle has a corollary: when excessively detailed parts do not contain much contribution, the structure lacks balance.

- A secondary part may be too detailed. Simplify or move details to appendix, footnotes, or supplementary material.[2]

---

[2] The supplementary material is not published. It is sent to the journal alongside your paper to enable the reviewer to check your data at a higher level of detail than that used in the paper, to provide additional proof which you left out, or to verify a formula which you could not expand in detail for lack of space and ease of reading.

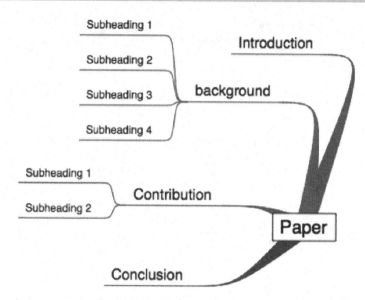

Figure ☞2
This structure points to one or several of the following problems: (a) background is too detailed; (b) contribution is small, therefore writer fills up paper with background; (c) Writer underestimates the knowledge level of the reader.

- The knowledge level of the reader is underestimated. Remove details and provide references to seminal papers and books (☞2).
- Subheadings are "sliced and diced" too small. When a section with only one or two short paragraphs has its own subheading, it should be merged with other sections.
- The top-level structure is not divided into enough parts. For example, the background section is merged with the introduction. As a result, many subheadings are necessary within the introduction. Add headings at the top-level of your structure to reduce the number of subheadings.
- The paper has a multifaceted contribution that requires a large background and an extensive structure. Rewrite it as several smaller papers (☞3).

Principle 2: Headings and subheadings detailing the contribution are grouped

When the branches containing the contribution are scattered instead of being grouped, the structure lacks singleness of focus (☞4).

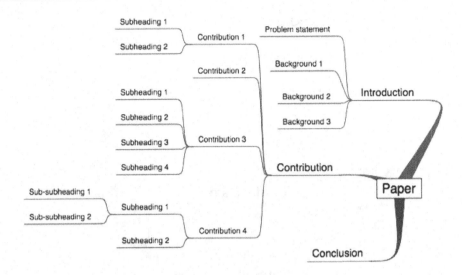

Figure ☞3
This structure points to one or several of the following problems: (a) the top-level structure has too few headings; (b) the contribution is too large for one single paper; (c) subheadings need to be merged.

- There may be more than one contribution in the paper; to make sure the paper is accepted, the writer stuffs his paper with contributions. The titles of such papers are very difficult to write! It is better to focus one paper around one contribution and have another paper (or letter) cover the additional contribution.
- The paper may not yet be ready for publication. The contribution is scattered in disconnected parts. The paper lacks unity and conciseness: repetition is unavoidable in such cases.
- The writer may be unable to identify the main contribution of the paper, or unable to establish a priority between major and minor contributions. The paper and its title probably lack focus.

Principle 3: Title words describing the contribution are repeated in the headings and subheadings of a structure

A structure disconnected from its title is either unhelpful or indicative of a wrong title. Since the role of a structure is to help the reader navigate inside your paper and identify where your contribution is located, a structure should have its headings and subheadings connected to the title (☞5).

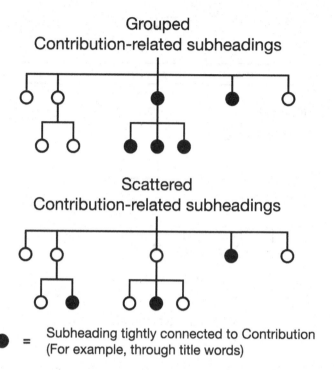

= Subheading tightly connected to Contribution (For example, through title words)

Figure ◄4
Grouped headings and subheadings show that the contribution is well identified and unique. When headings and subheadings covering the contribution in detail are dispersed throughout the structure, the structure has problems.

Let us apply the third principle to our sample structure and consider headings 2 and 3.

2. Mechanical factors underlying **slit** opening

Heading 2 contains '*slit*,' a title word found in the first part of the title describing the contribution. Heading 2 seems to be written for surgeons, whereas, heading 3 is definitely written for people on the IT side of mechanical engineering. Recall that the paper is published in the journal of Biomechanics, read by scientists from two different words — life sciences and engineering sciences — who may have difficulties understanding each other's work for lack of background knowledge. Therefore, it does make sense for the structure to address both types of readers.

Heading 2 takes us through the surgery steps of slit arteriotomy and the mechanically induced stresses and deformations observed during the surgery. At this point, surgeons stop reading but mechani-

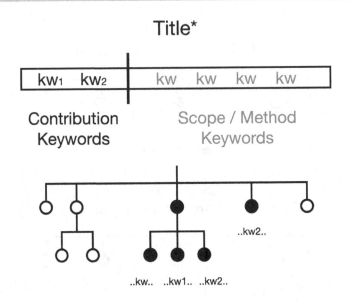

Title*

| kw₁ kw₂ | kw kw kw kw |

Contribution Keywords   Scope / Method Keywords

..kw2..

..kw.. ..kw1.. ..kw2..

* Title with no conjugated verb

Figure 5
Title keywords reappear in structure headings and subheadings to allow direct reader access to key parts of the paper.

cal engineers read on and find in heading 3 the model for the steps described in heading 2.

> 3.   Methodology for computer **simulation**
>   3.1   Reference configuration for the **finite element** model
>   3.2   Geometry details and boundary conditions of the **finite element** model in the reference configuration
>   3.3   Hyperelastic material for the arteries
>   3.4   **Simulation** procedure for the operation

Heading 3 and its four subheadings contain 'simulation' and 'finite element', two words located in the front part of the title (contribution part). They confirm that this heading, like heading 2, belongs to the contribution-related section of the paper. The author could have added 'nonlinear' to strengthen the coherence between title and structure. The specificity of the words in heading 3 and its subheadings conveys to the surgeons the message that this section of the paper is not written for them.

This third principle has a corollary: When headings and subheadings are disconnected from the title of a paper, the structure OR the title may be wrong.

- The structure reflects the contribution better than the title. For example, a structure where the word *'trajectory'* appears in three of the five headings, and yet does not even appear once in the title, betrays an imperfect title.
- The structure is too cryptic. Its headings and subheadings are too generic, brief, or tangential. They do not give enough information on the contents. Revise the structure and reconnect it to the title.
- The structure contains synonyms of the title words, or specific keywords replacing the title's generic keywords or even an acronym made of title words. Having lost homogeneity and coherence, the article is less clear. Return to the original keywords, or change the title to make it more specific.

Principle 4: A structure tells a story that is clear and complete in its broad lines

According to this fourth principle, someone unfamiliar with the domain of computer simulations should be able to see the logic of the story after reading the title, the abstract, and the successive headings and subheadings.

Is this story clear?

1. Introduction
2. Mechanical factors underlying slit opening
3. Methodology for computer simulation
   3.1 Reference configuration for the finite element model
   3.2 Geometry details and boundary conditions of the finite element model in the reference configuration
   3.3 Hyperelastic material for the arteries
   3.4 Simulation procedure for the operation
4. Results and discussion
5. Conclusions

Heading 2 takes the reader into the operating theater, to observe the surgeon cut and stitch the arteries. Under the sharp blade of the scalpel, the arteries open; they deform under the pressure of fingers and the pull of stitches. Once the surgery is completed, the reader can imagine the blood flowing through the arteries, opening the slit wider.

Heading 3 provides details on the simulation.

Subheading 3.1 defines the initial state of the simulated objects.

Subheading 3.2 gives details on the model parameters (arteries, slit) and defines their limits.

Subheading 3.3 describes how the arteries, key objects in the simulation, are modeled.

Subheading 3.4 makes the simulation steps correspond to the steps of the actual surgery.

The story is coherent with what the title announces, but it is incomplete. There is no link between the model and the result (elucidation). This could easily have been achieved by replacing the standard heading "*Results and discussion*" with a more informative heading such as "*Elucidation of the efficacy of slit arteriotomy*," thus establishing a direct connection between the model and its results, and encouraging the surgeons to read this section also.

The story may not be clear to surgeons unfamiliar with finite element modeling. This is clearly seen by the expert vocabulary used in headings 3.1 and 3.2. Six of their words (*reference, configuration, geometry, details, boundary, conditions*) are not even found in the abstract, whereas the words used by all other subheadings are found in the abstract.

The fourth principle has a corollary: when headings or subheadings reviewed in sequence tell a nonsensical story, the structure has holes or the title is wrong.

- The paper could be premature: its structure has not yet reached clarity. More work is needed until the structure falls into place.
- The story is nonsensical because it is not the story of the title, but another story. Change the title or rewrite the paper. You have the wrong face for the right body, or vice versa.
- The headings and subheadings are too concise or too cryptic — possibly because of the use of acronyms, synonyms, or highly specific keywords understood only by experts. Write more informative and understandable headings and subheadings.
- Key subheadings or headings are missing. Subdivide the obscure heading/subheading to reveal the missing connectors, or insert a new heading or subheading in the structure.

Syntactic Rules for Headings

Michael Alley, an author I designated as one of the giants who enabled me to see further afield in the world of scientific writing, advocates parallel syntax in structures. It does make the structure easier to read. Three styles are commonly found in headings and subheadings: Noun phrases such as *'Parameter determination'*, verb phrases such as *'Determining the parameters,'* and (in life science papers mostly) full sentences such as *'The parameters are determined statically'.* To help the reader rebuild a story from its structure, headings at the same indentation level or subheadings under the same heading should adopt a parallel syntax. In the sample structure, headings 2 and 3 are noun phrases. Within heading 3, all subheadings are also noun phrases.

In the following structure, the syntax is not parallel.

1. Introduction
2. Interference mechanism
3. Design rules
4. Proposing a solution
    4.1 Three layer prediction algorithm
        4.1.1 Algorithm classification
        4.1.2 Layer prediction comparison
5. Proposed recognition
6. Simulation studies
7. Discussion
8. Conclusions

This is not a good structure for many reasons. Focusing solely on the lack of consistency, one cannot miss the "one parent and only one child" problem: heading 4 has one subheading only 4.1 (no 4.2). The syntax also lacks consistency at the same heading level: headings 1, 2, 3, 5, 6, 7, and 8 are all single-noun phrases; but heading 4 starts with a present participle *'Proposing,'* thus breaking parallelism in syntax.

Purpose and Qualities of Structures

Purpose of the structure for the reader

- It makes navigation easy by providing direct access to parts of the paper.

- It helps the reader rapidly locate the section of the paper related to the author's contribution.
- It allows the reader to quickly grasp the main story of the paper by making a logical story out of the succession of headings and subheadings.
- It sets reading time expectations through the length and detail level of each section.

Purpose of the structure for the writer

- It reinforces the contribution by repeating Title/Abstract keywords in the headings or subheadings.
- It helps the writer divide the paper into informative, non-redundant, and logically connected sections that support the contribution.

Qualities of a structure

INFORMATIVE. No empty signposts should be found outside of the expected standard headings. Reading the structure headings and subheadings should be enough for readers to understand the title or to immediately identify a place of interest.

TIED TO TITLE AND ABSTRACT. Keywords from the title and abstract are also in the structure. They support the contribution.

LOGICAL. Going from one heading or subheading to the next, the non-expert reader sees the logic of the order chosen by the writer. There are no logical gaps.

CONSISTENT at the syntax level. Each parent heading has more than one child subheading. Syntax is parallel.

CLEAR AND CONCISE. Neither too detailed nor too condensed.

Here is a very simple and productive method to ascertain the quality of your structure. "Flatten" your structure on a blank piece of paper. By this I mean, write the title at the top of the page, and then write ALL headings and subheadings in the order they appear in your paper. Once done, underline the words structure and title have in common. Do you detect any discrepancy here? Are contribution keywords from the title missing in the structure? Is it because you are using synonyms? Change your structure to bring it closer to the title. Are you using acronyms that are defined in the text and make your headings obscure? If so, remove them or give their full name. Do you have frequent structure keywords which are absent from your title? Shouldn't they be part of the title? Could it be that your structure is right and your title is wrong? If so, change the title.

Once you have examined how well the structure matches your title, have someone else read your flattened structure and explain to you what he or she thinks your paper contains. The less this person knows about your work, the better. Ask this person if the logic is visible in the succession of headings or subheadings. If the person is largely puzzled, you are not quite ready to publish yet. Rework your structure and your paper.

When the story is clear, give a quick syntactic check. Is the syntax of your headings parallel? Are subheadings orphans?

When the volunteer reviewer asks questions, do not start explaining! Remember that you will not be there to explain your structure to your readers once your paper is published. Just take note of the questions, and adjust your structure or title accordingly.

Structure Q&A

Q: The structure is like an outline. Shouldn't we start a paper by creating its outline first?

A: Some writers use the structure as a framework for writing. They create the structure in bullet point form or in outline form, and then expand the points. This method has value. It gives focus to a paper. If the story flows well at the structure level, it will probably flow well at the detailed level also. The writer may still change the structure later, but it will mostly be to refine the headings or to create more subheadings, not to totally restructure the flow of the paper. Actually, as a side comment, when asked if they know that Microsoft Word includes an outlining program, most scientists say that they are not aware of such a capability! It is right there at your fingertips, an item under the *view* menu. It comes with its own set of tools. Play with it. I discovered outliners on the Apple II (a program called "*More*"). Life has never been the same for me since then. Even the first edition of this book started its life as an 11 page outline!

Q: To keep my headings short, can I use acronyms?

A: Using acronyms in headings or subheadings, unless they are well-known, is troublesome for the majority of readers. Obscure acronyms prevent readers from making a story out of the subheadings. But this is true also of highly specific expert words. Write your headings and subheadings for non-experts. And if you absolutely need to use an expert word in a subheading, define it right away in the first sentence that follows that subheading. Doing so respects the principle of just-in-time background.

Q: What makes a good subheading?

A: A good subheading is an informative noun phrase whose keywords appear with high frequency in the paragraphs that follow it. It fits neatly and logically in between other subheadings. And it is closely related to the contribution, often by including words from the title and the abstract. A bad subheading contains acronyms, synonyms of the title keywords, or specific instances of title keywords (for example the title contains 'memory,' and the subheading contains 'register,' a specific keyword not even used in the abstract). The computer science expert knows the connection between memory and register, but the non-expert with a knowledge gap cannot make the connection.

Q: How many paragraphs come under a subheading?

A: That will depend on the length of your paragraphs, and on the flow of your paper. For example, if you describe a succession of steps, but each step is covered in one paragraph of two to four sentences, it may make more sense to group or collapse several steps under a more comprehensive subheading name. Let the specificity of the keywords in the subheading guide you. If the subheading is so specific that its keywords do not even appear in the abstract, reconsider its usefulness or its wording.

Q: My journal dictates the labeling of the subheadings! Can I change them?

A: To enable the reader to reproduce your results, some journals (chemistry or life science journals come to mind) demand that you follow their imposed structure. If a journal dictates the number and types of subheadings, you have no choice but to comply.

Q: I wrote a letter, not a paper. My letter does not have headings, should I include some?

A: Not all papers have an explicit structure. When the paper is short, a letter for example, there is no need to add a structure: the structure is implicit. The "Introduction" heading is absent, but the first paragraph of the letter introduces. The "conclusions" heading is absent, but the last paragraph concludes. However, if you want to benefit from the tests and the metric in this chapter, I recommend that you artificially create and write down such a structure based on the contents of your paragraphs, even though it won't be part of your letter.

Q: My structure has standard headings (intro, results, ...). Should I include more subheadings?

A: If your paper has enough pages, you should at least add two subheadings (or more) under the heading that contains the bulk of your contribution. It facilitates reading.

Q: I have read the structure of my friend's paper, and found it very difficult to connect with the title of his paper. What is wrong: the title, or the structure?

A: It could be both. If you were not able to retell the story of the title just by browsing the headings and subheadings of your friend's paper, you may trace the problem back to five main sources: (1) synonyms (title and structure use different keywords meaning the same thing), (2) Acronyms (defined elsewhere), (3) uninformative headings/subheadings (hollow markers such as *experiment*), (4) logical gaps, or (5) unintelligible expert keywords.

Q: Does every single word in the structure need to be taken from the title or abstract?

A: Unless these words are generic words known by all, including well known acronyms, the words in your top-level subheadings need to be taken from the title and abstract. At deeper subheading levels (for example 1.1.2), you can choose any word you like because readers are unlikely to read these subheadings during their first discovery of the paper.

Q: My paper is a review. Do the principles detailed here also apply to this kind of paper?

A: A review paper is quite different, but principle 4 (the story line) should still be respected.

Structure Metrics

✓(+) All contribution-related search keywords in the title are also in the headings/subheadings.

✓(+) The contribution is grouped under successive headings.

✓(+) The structure contains informative subheadings

✓(+) Not one heading/subheading could change place without compromising the structure of the paper.

✓(+) In top level subheadings, the structure does not contain acronyms, synonyms, or keywords understood by experts only.

✓(+) The majority of your structure words are in the abstract.

✓(+) The structure does not contain orphan headings or subheadings.

✓(−) Contribution-related search keywords from the title are missing in your structure.

✓(−) The contribution is scattered throughout the structure.

✓(−) The structure does not contain any subheadings, or any informative headings or subheadings.

✓(−) Headings/subheadings could change place without compromising the structure of the paper.

✓(−) In top level subheadings, the structure contains acronyms, synonyms, or specific expert keywords.

✓(−) Less than 50% of your informative structure words are in the abstract.

✓(−) The structure contains orphan headings or subheadings.

AND NOW FOR THE BONUS POINTS:

✓(+++) Even a non-expert could figure out the title of your paper, just by reading the structure.

Chapter 16

Introduction: The Hands
of Your Paper

Extended hands welcome and invite to enter. They guide someone unfamiliar with a new place. The introduction of a paper plays a similar role. It greets, provides guidance, and introduces a topic not familiar to the reader. *Hands point to something worthy of attention, and invite the eyes to follow.* The introduction also points to the related works of other scientists and to your contribution.

For many, the introduction is a necessary evil, something more difficult to write than the methodology or results section. Therefore, to ease the burden, the scientist usually keeps it brief. Alas, brevity is only appreciated by the few experts in the field already familiar with the introduction material. The *many* readers (a reasonable 40%) with a significant knowledge gap will not be satisfied. Reviewers could even be among them. Ask a reviewer how many papers he or she accepts to review without full expertise in the matter being reviewed, and be ready to be surprised! Therefore, write an introduction that bridges their knowledge gap, otherwise they may not be able to evaluate your paper correctly. Remember that they have veto power over the selection of your paper for publication.

Writing Against All Odds

Vladimir, rejected paper in hand, was reading the online comments from the three people who had reviewed his paper. From the remarks, Vladimir could tell that two of them were knowledgeable in his field. It was the remarks of the third reviewer that bothered him. Clearly, that was not someone from his field. He had presumed all of them would be experts.

Popov, his supervisor, was passing by. Vladimir called him.

"Hey, boss, aren't you supposed to be an expert in the field before you are invited to review papers?"

"Often the case. May be 80% of the time."

"You mean you have reviewed papers that are outside of your direct field of expertise?"

"Sorry, Vlad, no time to chat, have to run off to a meeting."

Left alone, Vladimir did a quick mental calculation. Let's see. If I have three reviewers, what is the probability that I get at least one reviewer who is not an expert if each reviewer's probability to be an expert is 80%... Hum, that would be... point eight times point eight, that's point sixty-four times point eight again, and that's...sixty times eight, four hundred and eighty, plus four times eight, thirty-two, that makes ...51.2% chances that all reviewers are experts. So I have one chance in two that at least one of my three reviewers is not an expert!

On his way back shortly after, Popov stopped in front of Vladimir's cubicle. "Meeting cancelled," he said. What is it you wanted to know?"

"Are you sure about your 80%?" asked Vladimir.

"80% what?" His boss had already forgotten.

"80% chance to be a direct expert in the topic of the paper you are asked to review."

"That sounds about right. It depends on the journal of course. For large journals, the percentage may be even lower."

"And out of three reviewers, how many have to recommend your paper for it to have a chance to be published?"

"Again, that will depend on the journal, but for prestigious journals, I'd say, all three."

"Great! Vladimir mumbled. That means that I have to write my paper to be understood by amateurs!"

"Are you calling me an amateur, Vladimir, ... because in front of you stands the amateur who has reviewed many papers where he did not consider himself to be an expert in everything that was written."

"Oh, no, Boss... You're a pro, right John?"

From across the cubicle partition, John's voice shouted:

"A real pro, Boss!"

The Introduction Starts Fast and Finishes Strong

The introduction engages. Do not keep the reader's brain idling for too long. Your reader is eager to go places. How do you know for sure? Well, the impatient reader having gone past two filters, the title (coarse grain) and the abstract (fine grain), decided to click on the download button. Interest must be high! Do not bore or delay the reader. Do not disappoint with a false start that slows the pace of your introduction.

The vacuous false start

> In the age of genomes, large-scale data are produced by numerous scientific groups all over the world.

> Significant progress in the chemical sciences in general, and crystallography in particular, is often highly dependent on extracting meaningful knowledge from a considerable amount of experimental data. Such experimental measurements are made on a wide range of instruments.

> Because of the long-term trend towards smaller and smaller consumer goods, the need for the manufacture of micro components is growing.

Was there anything in these examples you did not already know? Catch and ruthlessly exterminate these cold starts, these hollow statements where the writer warms up with a few brain push-ups before getting down to the matter at hand. By the way, if you are such a writer, do not feel bad. Many of us are. It takes a little while for our brain to output coherent and interesting words when we face a blank page or a blank screen! Since we have a propensity to do brain push-ups at such times, let's make sure to remove the sentences they produce.

Here is another type of false start. At first glance, there seems to be nothing wrong. After all, the writer tries to conjure up excitement by showing how massively important the current problem is.

The considerable false start

> There has been a surge, in recent times, towards the increasing use of ...

> There has been considerable interest in recent years in this technology, and, as trends indicate, it is expected to show continuing growth over the next decade ...

In this type of false start, the author sees the heat of a research field as being sufficient to warm the reader to his contribution. The words used to raise the temperature are *'exponential,' 'considerable,' 'surge,' 'growing,' 'increasing,'* and other ballooning words. They lead an important class of readers, the reviewers, to suspect a "me-too" paper. There may have been a recent surge but the writer is obviously running behind the pack of researchers who created the surge. People who are pioneering research make the future, they don't catch up with the past!

Reviewers may also suspect that the writer is attempting to influence their judgment by equating importance of problem to significance of solution. If many people consider the problem important, does that make the contribution an important one? There is no necessary cause and effect relationship here. If I shake hands with a nobel prize winner, does that make me a great scientist? If my neighbor has been arrested for robbery, does that make me a felon?

The right start

It is best to start with what readers expect: they want more details. Your title is sketchy. It draws the silhouette of a face. Your abstract puts a bright but narrow frontal spotlight on the face, giving it a flat, even look and making everything else recede into the shadows. The introduction is the soft filling light. It adds dimension, softens the shadows, and reveals the background. To use a metaphor familiar to photographers, the abstract privileges speed of capture whereas the introduction emphasizes depth of field.

At the start of the introduction, since the face is still in full sight of a reader, it is best to remain close to the title and frame it in context. The context, be it historical, geographical, and even lexical (definitions), should not lose its relationship with the face. Returning to the photo metaphor, the reader should constantly see the face in the viewfinder even when the perspective widens. Too many papers paint a landscape so vast in the introduction that the reader is unable to place the title in it.

Here is a fast and focused start combining definition and historical perspective.

Name Entity Recognition (NER), an information extraction task, automatically identifies named entities and classifies them into predefined classes. NER has been applied to Newswires successfully [references]. Today, researchers are adapting NER systems to extract biomedical named entities — protein, gene or virus — [more references] for applications such as automatic build of biomedical databases. Despite early promising results, NER's ability to apply to such entities has fallen short of people's expectations.

After reading this paragraph, the reader expects the writer to explain why success is limited, and to bring an answer to the main question: what adaptations to the original NER would enable biomedical named entities to be extracted more successfully.

The dead end

At the Doorstep — Rejected!

Steve Wilkinson is an insurance agent for a large insurance company. He is also Vladimir's neighbor. For weeks, he has been chatting across the fence with Vladimir's wife, Ruslana (about insurance of course), while she prunes her rose bushes in the garden. She finally tells Vladimir who, being quite good at maths and curious to find out about the benefits of insuring his family, decides to accept his neighbor's invitation to hear about his company's insurance products. A date is fixed: next Thursday after work.

Thursday evening. Vladimir shows up at Mr. Wilkinson's front door and rings the bell. Steve goes to open the door, and immediately enters into a monologue at the doorstep before letting Vladimir in.

"Recently, an increasing number of people buy insurance because of global warming. Global warming may lead to life and property threatening weather. AIE, and PRUDENTA offer insurance schemes in this domain. A different scheme is examined today. After four years of insurance premiums, the GLObal WARming BLanket Insurance Scheme, GLOWARBLIS, provides a 5.6% yearly interest on the insurance premiums, assuming the global warming index remains stable four years in a row. A 1-point fluctuation in the global warming index during this period will decrease the yearly interest rate by a corresponding amount. After entering this house, go through the corridor leading to the living room where the prospectus will be examined. Following this, proceed to the office to discuss the signing of the contract. After signature, return to this door to leave with a package presenting other insurance schemes from the company."

Table-of-content type endings have no place in an introduction where readers can just flip a few pages and discover the whole structure by rapidly scanning the headings and subheadings. Therefore, don't end your introduction as seen below.

The rest of this paper is organized as follows. Section 2 discusses related work. Section 3 presents the technology and shows how our approach is conducted using our scheme. Section 4 presents the results of our experiments and shows how the efficiency and accuracy of our approach compare with others. Finally, we offer our conclusions and discuss limitations.

The rest of this paper is organized as follows. Section 2 describes some related works, in particular similar work that has been done. Following that, the proposed approaches are discussed in section 3, with the implementation details being discussed in section 4. Section 5 evaluates the performance and compares the proposed approaches to a baseline model. Finally, we draw conclusions and outline future work in section 6.

When a place is small and more or less standard — for example the house of the insurance agent in the Vladimir story — there is no need to describe the various rooms and tasks that lay ahead. But when the place is large and not standard (the White House in Washington, the Imperial Palace in Beijing, or the Versailles Castle in France), it is wise for the guide to give a brief overview of the visit (schedule, route followed…) before entering. A Ph.D thesis, or a book, is large enough to deserve an introduction that helps the reader anticipate what are the main parts and what they accomplish.

The strong finish

> ### At the Doorstep
>
> *Steve Wilkinson is an insurance agent for a large insurance company. He is also Vladimir's neighbor. For weeks, he has been chatting across the fence with Vladimir's wife, Ruslana (about insurance of course), while she prunes her rose bushes in the garden. She finally tells Vladimir who, being quite good at maths, and curious to find out about the benefits of insuring his family, decides to accept his neighbor's invitation to hear about his company's insurance products. A date is fixed: next Thursday after work.*
>
> *Thursday evening. Steve sees Vladimir approaching his house. He goes to open the door even before Vladimir has a chance to ring the doorbell, and warmly greets his neighbor.*
>
> *"Vladimir, I'm so glad you could come. Today is a day you will remember as the day you found peace by sheltering your family from the financial worries that global warming brings with its devastating tornadoes and countless other financial calamities. Come in, come in."*

The best ending to an introduction is in this second Hollywood-approved scenario: Vladimir is told about the outcome of signing the insurance contract, and then moves right in. End your introduction with the expected post-contribution outcome of your research to keep reader motivation high.

 Read the first paragraph of your introduction. Is it vacuous or considerable? If it is, delete it. Is the last paragraph redundant with the structure? If it is, delete it.

The Introduction Answers Key Reader Questions

What is the main question of your paper? It is the question that you answer by stating your contribution. If you cannot phrase this question, you are not ready to write your paper because you do not yet have a clear idea of your contribution. To help you determine the main question, practice on the following familiar titles:

"Non-linear Finite Element Simulation to Elucidate the Efficacy of Slit Arteriotomy for End-to-side Arterial Anastomosis in Microsurgery"

Main question:

Having observed the efficacy of slit-arteriotomy in terms of blood flow and robustness of anastomosis for patients of various age and blood pressure conditions, can this efficacy be explained by modeling the mechanical stresses sustained by the arteries during and after the operation?

"Energy-Efficient Data Gathering in Large Wireless Sensor Networks"

Main question:

How can a sensor node be chosen to forward data in a large wireless sensor network so that total energy consumption for data forwarding in the network is minimized?

 Read your title and abstract. Write the main question they answer. Is that question clearly stated in your introduction? If there is more than one question, you may have a paper with multiple contributions, and possibly a paper that could be divided into multiple papers. Or, you may not yet clearly perceive your contribution.

As soon as you identify the main question answered by your paper, include it in your introduction. And here is an "Oh, by the way" remark that may turn out to be the only help your introduction needs to be close to perfect. ***"Oh, by the way, have you thought of putting this question to the reader in visual form?"*** As you know by now, nothing is better than a visual to demonstrate and convince. So make sure you spend the time to clarify first in your mind for yourself, and then on paper for the reader, that "mother of all questions".

From that main question comes a mother lode of other questions: Why this? Why now? Why this way? Why should the reader care (how is your work relevant to the reader's needs)?

> ### The Questionable Cake
>
> *That afternoon, Vladimir Toldoff received a call from his wife Ruslana as he was finishing an experiment in the lab.*
>
> *"I am coming with a cake, the cake knife, and enough plates and cutlery for four or five," she announced.*
>
> *He answered, "What? Wait! First, what is the occasion? And why now, can't it wait until tonight? And by the way, what cake is it, and why do you want to cut it in the lab? You know that crumbs are not welcomed here."*
>
> *The rapid-fire questions did not faze Ruslana. She knew her Vladimir. A full-fledged scientist. She paused and rephrased his questions succinctly.*
>
> *"All right, let me see. You would like to know why a cake, why eat it now, why its mouth-watering taste should make you shout 'Darling come right away,' and why I should slice it in the lab instead of at home, am I right?"*
>
> *Vladimir grinned. He was quite impressed with his wife's listening skills.*
>
> *"That's right, Mrs Toldoff," he responded.*
>
> *Ruslana then uttered four words that had him shout for joy: "Your Medovik birthday cake".*

Through these *why* questions, the reader expects you to justify your research goals, your approach, as well as the timeliness and value of your contribution.

The following example taken from a life science paper illustrates how the writer answered these questions; it contains acronyms or jargon you may be unfamiliar with, but that should not deter you! The paper's title is the following:

"*A gene expression-based method to diagnose **clinically distinct** subgroups of Diffuse Large B Cell Lymphoma*" (DLBCL)

Why now? In this case, because recent studies present diverging results.

"*We were curious to see whether we could resolve the discrepancy between these gene profiling studies by using our current understanding of the gene differences between GCB and ABC DLBCL.*"[1]

---

[1] George Wright, Bruce Tan, Andreas Rosenwald, Elain Hurt, Adrian Wiestner, and Louis M. Staudt, a gene expression-based method to diagnose clinically distinct

Why this? In this case, because there is a need for labs to deliver a coherent clinical diagnosis regardless of the platform used to measure gene expression.

> "As was pointed out (3), it is a challenging task to compare the results of these profiling studies because they used different microarray platforms that were only partially overlapping in gene composition. Notably, the Affymetrix arrays lacked many of the genes on the lymphochip microarrays …"[2]

Why this way? In this case, because the method works on different microarray platforms.

> "For this reason we developed a classification method that focuses on those genes that discriminate the Germinal Centre B-cell like (GCB) and the Activated B-Cell like (ABC) Diffuse Large B Cell Lymphoma (DLBCL) subgroups with highest significance."[3]

Why should the reader care? In this case, because it predicts survival regardless of the experimental platform.

> "Our method does not merely assign a tumor to a DLBCL subgroup but also estimates the probability that the tumor belongs to the subgroup. We demonstrate that this method is capable of classifying a tumor irrespective of which experimental platform is used to measure gene expression. The GCB and ABC DLBCL subgroups defined by using this predictor have significantly different survival rates after chemotherapy."[4]

To these reader questions, the **reviewer** adds other questions. Even though they overlap, they differ in important ways.

1) Is the problem clear, and is solving it useful?
2) Is the solution novel, and is it much better than others?

Therefore, you should have both reader and reviewer in mind when you write your introduction. It is up to you to convince them that the problem is real, and that your solution is original and useful.

Subgroups of diffuse large B Cell Lymphoma, Proceedings of the national Academy of Sciences, August 19, 2003 Vol. 100 No.17, www.pnas.org, copyright 2003 National Academy of Sciences, U.S.A.

[2] ibid.
[3] ibid.
[4] ibid.

The Introduction Frames Through Scope and Definitions

The writer's intellectual honesty is demonstrated in many ways. One of them is a clear and honest description of the scope. Readers need to know the scope of your work because they **want** to benefit from it, and therefore need to evaluate how well your solution might work on their problems. If the scope of your solution covers their area of need, they will be satisfied. If it does not, at least they will know why, and they may even be encouraged to extend your work to solve their problems. Either way, your work will have been helpful.

Scope

In essence, the scope or frame of your contribution is set by the methodology, the data, the time frame, and the application field. Establishing a frame around problem and solution enables you to claim with some authority and assurance that your solution is "good" within that frame. Some writers leave the framing until later in the paper. I believe that a reader informed on the scope early is better than a reader disappointed by late disclosures restricting the applicability and the value of your work. Therefore, establish the scope early in your paper.

Use the answer to the "Why this way" question to make the scope more precise while enhancing the attractiveness of your paper and highlighting its novelty.

> Our method does not need a kernel function, nor does it require remapping from a lower dimension space to a higher dimension space.

> Our dithering algorithm does not make any assumption on the resolution of pictures nor does it make any assumption on the color depth of the pixels.

Not all assumptions in your paper affect the scope. These assumptions are best mentioned in the paragraphs to which they apply (often in the methodology section). Justify their use, or give a measure of their impact on your results, as in the following three examples: (1) *Using the same assumption as in [7], we consider that...; (2) Without loss of generality, it is also assumed that ...; (3) Because we assume that the event is slow varying, it is reasonable to update the information on event allocation after all other steps.*

Define

Another way of framing is by defining. In the following example, the authors define what an *'effective'* solution is. They do not let readers decide on their own what *'effective'* means.

> *"An effective signature scheme should have the following desirable features:*
>
> - *Security: the ability to prevent attacked images from passing verification;*
> - *Robustness: the ability to tolerate ...*
> - *Integrity: the ability to integrate ...*
> - *..."*[5]

When you define, you frame by restricting the meaning of the words to your definition. Demonstrating a solution is *effective* because it fulfills predefined criteria is easier than demonstrating a solution is *effective* when the evaluation criteria are up to the reader.

Scope and definitions ground the readers' expectations and set strong foundations for the credibility of your results. In the next chapters two other credibility-enhancers will be considered: citations and precision.

The Introduction Is a Personal Active Story

Personal

Your contribution is woven into the work of other scientists (the referenced related work). Clearly identify what is yours and what belongs to others by using the personal pronouns *'we'* or *'our'*.

Naturally, using personal pronouns changes the writing style. Thankfully, there is not **one** scientific writing style, as many believe. There are two: the personal and the impersonal style. The introduction reinforces the motivation of the reader to read the rest of the paper, by being low on difficult content, and high in active voice and personal pronouns.

---

[5] Shuiming Ye, Zhicheng Zhou, Qibin Sun, Ee-chien Chang and Qi Tian, a quantization-based image authentication system. ICICS-PCM, Vol. 2: 955–959. © I.E.E.E 2003

The Story of Vladimir Toldoff

"Vladimir!"

The finger of Popov, his supervisor, is pointing at a word in the third paragraph of Vladimir's revised introduction.

"You cannot use 'we' in a scientific paper. You are a scientist, Vlad, not Tolstoy. A scientist's work speaks for itself. A scientist disappears behind his work. You don't matter Vlad. 'The data suggest'… you cannot write "our data". It's THE data, Vlad. Data do not belong to you. They belong to Science! They speak for themselves, objectively. You, on the other hand, will only mess things up, introduce bias and subjectivity. No Vlad, I'm telling you: stick to the scientific traditions of your forefathers. Turn the sentences around so that you, the scientist, become invisible. Write everything in the passive voice. Am I clear?"

"Crystal," Vladimir responds, "But I was only taking the reviewer's comments into account."

And with that, he hands out a printed copy of the comments he had received from the reviewer.

Popov grabs the paper.

"What kind of nonsense is this?", he says.

(reading the letter aloud)

…Your related work section is not clear. You write, "The data suggest". Which data? Is it the data of [3], or is it your data? If you want me to assess your contribution fairly, you should make clear what YOUR work is and what the work of others is. Therefore, if it is your data, then write, "our data suggest". Also, if I may make a suggestion, I feel that your introduction is somewhat impersonal and hard to read. You could improve it by using more active verbs. That would make reading easier….

"Ah, Vladimir! No doubt, this comes from a junior reviewer. What is happening to Science!"

Even if you are the first author of the paper, research is, for the most part, a collaborative effort. It is therefore appropriate to use personal pronouns such as *we* or *our*, leaving the personal pronoun *I* for later in your professorial career when you write papers alone. Some of the old guard object to the use of personal pronouns, and advocate the passive voice to give the paper a more authoritative disembodied voice, but I will argue that it is difficult to welcome a reader into the body of your research this way.

> ### The Story of the Passive Lover
>
> *Imagine yourself at the doorstep of your loved one. You are clutching, somewhat nervously, a beautiful bouquet of fragrant roses behind your back. You ring the doorbell. As your loved one opens the door and gives you a beaming smile, you hand out the bouquet of flowers and utter these immortal words:*
>
> *"You are loved by me."*
> *What do you think happens next?*
> *(a) You eat the flowers; or*
> *(b) You ring the doorbell again and say the same thing, this time, using the active voice.*

Active story

The introduction of your paper helps guide the reader into the story of your research. But to be engaging, stories have to have certain qualities! As readers, we are interested in knowing WHO is doing WHAT. The active voice forces this sentence structure to always be present, as in *we considered all available options*. The passive voice allows for the subject to be hidden, as in *all available options were considered*. They were considered, yes, but by who? Stories without actors are boring, and you can't afford to start boring your reader in the introduction.

> *"**We** were **curious** to see whether **we** could resolve the discrepancy between these gene profiling studies by using **our** current understanding of the gene differences between GCB and ABC DLBCL."[6]*

See how abstract and introduction differ in writing style.

Abstract

> *"The GCB and ABC DLBCL subgroups identified in this data set had **significantly different 5-yr survival rates after the multiagent chemotherapy (62% vs. 26%; P = 0.0051),** in accord with analyses of*

---

[6] George Wright, Bruce Tan, Andreas Rosenwald, Elain Hurt, Adrian Wiestner, and Louis M. Staudt, a gene expression-based method to diagnose clinically distinct subgroups of diffuse large B Cell Lymphoma, Proceedings of the national Academy of Sciences, August 19, 2003 Vol. 100 No.17, www.pnas.org, copyright 2003 National Academy of Sciences, U.S.A.

*other DLBCL cohorts. These results demonstrate the ability of **this gene expression-based predictor** to classify DLBCLs into biologically and clinically distinct subgroups irrespective of the method used to measure gene expression."*[7]

Introduction

*"**We** demonstrate that this method is capable of classifying a tumour irrespective of which experimental platform is used to measure gene expression. The GCB and ABC DLBCL subgroups defined by using **this predictor** have **significantly different survival rates after chemotherapy**."*[8]

The abstract is more precise than the introduction when it comes to the key numerical results. But, the factual abstract does not tell a personal story: *'these results demonstrate'* is impersonal whereas *'We demonstrate'* is active and personal. The passive voice is quite acceptable in the rest of your paper where knowing who does what matters less. But in the introduction, the active voice rules!

Did you use the pronoun 'we'? Did you answer all the whys? Identify where each why is answered. Do you bridge the knowledge gap with some definitions? Did you build interest in the story? If not, why not? Did you cut and paste text between your abstract and your introduction? If you did, rewrite. To identify whether you adequately scoped your problem and solution, simply underline the sentences that deal with scope, assumptions and limitations. Are there enough of them? Are they at their proper place? Finally, did you mix introduction with technical background? If you did, place your technical background after the introduction under a separate heading. The introduction captures the mind; the technical background fills it. These two functions are best kept separate.

---

[7] *ibid.*
[8] *ibid.*

Chapter 17

Introduction Part II: Popular Traps

The introduction helps the reader understand the context from which your research originated. Do you borrow or adapt the work of other scientists to reach your objective, or do you follow a completely different research path? The reader wants to know. Positioning your work on the research landscape is a perilous exercise because it is tempting to justify your choices by criticizing the work of others. It is also tempting **not** to compare your work with the work of others so as not to upset anyone. It is equally tempting to borrow other people's ideas and conveniently forget to tell the reader where these ideas came from. Temptation is plaguing the scientist writing a paper simply because the stakes are high: publish **or _perish_**.

Five traps are laid in the path of writers: the trap of the story plot, the trap of plagiarism, the trap of references, the trap of imprecision, and the trap of judgmental words.

TRAP I — The Trap of the Story Plot

The introduction tells the personal story of your research. All good stories have a story plot to make them interesting and clear.

A story

I'm so excited to share this great story with you. My father [1] is on the front lawn cleaning the lawn mower. My sister [2] is in the kitchen making a cake. My mum [3] has gone shopping, and I am playing my electric guitar in my bedroom.

Do you like my story? No? It is a great story. What's that you're saying? My story has no plot? Of course there is a plot! See, it describes my family's activities, starting with my father. We all have something in common: family ties, living under one roof...

If this story left you cold, the analogous story found in scientific papers will also leave the reader cold. In short, the story says: in this domain, this particular researcher did this; that research lab did that; in Finland, this other researcher is doing something else; and I'm doing this particular thing. The problem with this type of story is that the relationship between their work and your work is not stated. In symbolic graphical form, the story plot would look like ☛1a. The pieces are juxtaposed, not linked.

Figure ☛1a
All story elements are juxtaposed, disconnected: father, mother, sister, brother.

Contrast the first story with this one:

> *I'm so excited. I'm going to tell you a great story. My father [1] is on the front lawn cleaning the lawn mower. And do you know what that means? Trouble! He hates it. He wants everyone to help bring this or bring that to make himself feel less miserable. When that happens, we all run away, not because we refuse to help him, but because he wants us to stand there and watch idly while he works. So my sister [2] takes refuge in the kitchen and plunges her hands in flour to slowly make a cake. My mum [3] suddenly discovers that she is missing parsley and rushes to the supermarket for an hour or so. As for me, I escape to the upstairs bedroom and play my electric guitar with the volume cranked up to rock concert levels.*

The difference is striking, isn't it? In an interesting story plot, all parts are connected as in ☛1b.

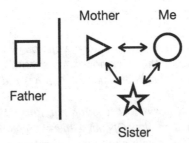

Figure ☛1b
Three story elements (mother, sister, me) share a common bond. This bond isolates the square element (father).

Here is a second story based on a story plot often found in scientific papers.

> ### A terrible story
>
> *I'm so excited. I'm going to tell you my second best story. A red Ferrari [1] would take me to the house of Vladimir Toldoff in 5 hours. It is fast. However, it is very expensive [2,3,4]. A red bicycle [5] is much less expensive and quite convenient for short trips. So, if Vladimir Toldoff comes to live near my house, it will be quite cost effective [6]. However, bicycles require accessories like mudguards or bicycle clips [7] to keep trousers clean. Red athletic shoes [8], however, require no accessories, and are as good a solution as a bicycle to travel over short distances [9]. However, their look is easily degraded [10] by adverse weather. On the other hand, brownish open plastic sandals [11] do not have any of the previous problems: they are cheap, weatherproof, convenient, and require no accessories. Furthermore, they are easily cleaned, and fast to put on. However, contrary to the Ferrari, they reflect poorly on the status [12] of their owner. Therefore, I am working on a framework to integrate self-awareness into means of transportation, and will validate it through the popular SIMs2 simulation package.*

Yes, I have exaggerated (only a little), but you get the point. The *however* plot, after taking readers through a series of sharp *however* turns, completely loses and confuses them. The seemingly logical connection between the elements is tenuous, as in ☛2.

Figure ☛2
4 shapes: a sun, a star, a cross, and an ellipse. The first element is compared with the second, the second with the third, and so on. At the end, the final element is connected back to the original element, thus completing the loop. Yet, the sun is never compared with the cross and the star is never compared with the ellipse. For the chain of comparisons to be meaningful, the comparison criteria must be identical for all elements, and all elements must be compared.

On the way to the last proposal (the writer's contribution), a long list of disconnected advantages and disadvantages is given; by the time readers get to the end of the list, they innocently (and wrongly) assume that the final solution will provide all the advantages and none of the disadvantages of the previous solutions. Unfortunately, the comparison criteria continuously vary, and therefore, nothing is really comparable.

Both plots, the juxtaposed story plot and the meandering story plot, are often found because they are convenient from a writer's perspective. There is no need to spend hours reading the papers referenced, reading their titles is enough (and you can find these in the list of references of other papers anyway), at a push the writer will read their abstracts, no more.

Are there better plots? Assuredly, but giving examples would fill too many pages, so here they are in schematic form ➛3.

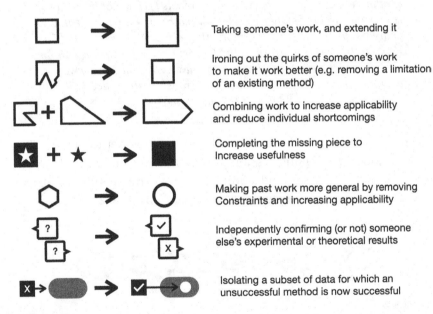

Figure ➛3
Various schematic story plots that work.

I have found that a popular movie plot is also useful in scientific writing. The author shows you how the story ends even before it starts. When readers have the full picture, they are better able to situate your work in it. They understand how and with whose help you will achieve your results. In addition, they know the scope of your work and are clear about the future work. Graphically, it is represented in ➛4.

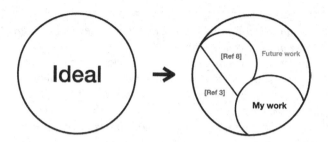

Figure ☛4
First the ideal system or solution is depicted (here represented by the circle). Then the story tells how this ideal picture comes together: what the author contributes, what others have already contributed [referenced work], and what remains an open field of research (future works). Everything is clear, everything fits nicely, and the reader is more easily convinced of the worth of your contribution.

 Identify your story plot. Does it look like a "however" meander or a series of juxtaposed disconnected elements? Is your story easy to follow? Does it flow logically: from past to recent, from general to specific, from specific to general, from primitive to sophisticated, from static to dynamic, from problem to solution, or from one element in a sequence to the next in line?

TRAP 2 — The Trap of Plagiarism

Plagiarism exists when someone else's words are found in your paper without proper quotes **and** references. Senior researchers, whose names often appear as the third or fourth author in a paper, do not need to be told. Their reputation is at stake. They know only too well the hefty price one pays when caught. They have heard the tale of the faculty dean high up in the research ladder who had to resign because someone found out that he had plagiarized in a paper written 20 years earlier when he was still a junior researcher.

> Vladimir Toldoff told off again
>
> *"Vladimir!"*
> *Vladimir's supervisor Popov points his finger to a sentence in the third paragraph of Vlad's introduction in the paper published three months earlier in a good journal.*
> *"Yes, anything wrong?"*

> *"The English in this paragraph about Leontiev's algorithm is too good. These are not your sentences."*
>
> *"Um, let me see. Ah, yes, it is rather good, isn't it! I must have been in great shape that day. I remember noticing how well I had worded that paragraph when I cut and pasted it into my paper from my reading notes."*
>
> *"Would it be too much to ask you to bring your reading notes?"*
>
> *"You have access to them already. I emailed you the files after the review meeting last month."*
>
> *"Oh yes. That's right. Give me a moment… Ah! Here are your notes on Leontiev's work, and here is that sentence. Now let me retrieve Leontiev's paper from the electronic library. Just a minute. Here it is. Let me copy a sentence from your paragraph and do a string search on Leontiev's paper and…well, well, well! What do we have here?! An exact copy of the original!"*
>
> *"Oh NO!" Vladimir turns red. But he recovers quickly and smiles widely. "It's fine! Look! I put a reference to Leontiev's work right at the end of the paragraph. A reference is the same as a quote, isn't it? After all, Leontiev should be happy. I am increasing his citation count. He will not come and bother me by claiming that these words are his, not mine."*
>
> *Popov remains silent. He retrieves from the top of his in-tray basket what looks like an official letter and reads it out loud. "Dear Sir, one of my students brought to my attention that a certain Vladimir Toldoff who works in your research center has not had the courtesy to quote me in his recent paper, but instead claimed my words to be his (see paragraph three of his introduction). I am disappointed that a prestigious Institute like yours does not carefully check its papers before publication. I expect to receive from your Institute and from Mr Toldoff a letter of apology, with a copy forwarded to the editor of the journal. I hope this will be the last time such misconduct occurs.*
>
> *Signed: Professor Leontiev."*

Plagiarism comes in two flavors: unintentional and intentional.

Vladimir's story illustrates a case of accidental plagiarism. It is often due to a less than perfect process to collect and annotate background material. Keeping relevant documentation about the information source when capturing information electronically is simply good practice. If you are unsure that a sentence is yours, just cut and paste it into the Google search window to see if it belongs to someone else. Another source of accidental plagiarism comes from writing a paragraph too soon after reading related ideas in someone's paper. What you read is still undigested, unprocessed, raw; parts of the sentences may still be lingering in memory. There has been no integration, no transformation, and no enrichment. Unintentional plagiarism may also come from sheer ignorance of the concept. I remember reading a paper where one paragraph was so much better written than the others that I asked where it came from. The researcher said that she found the paragraph describing the Java function on a website and it was so well written, so concise, that she could not have written it better herself. She simply assumed that being freely accessible on the web made it legal to copy without quote and attribution.

Intentional plagiarism is stealing for gain. It could be to gain time (no time to write a paraphrase or to retrieve the original paper where the sentence comes from). It could be to gain consideration by enlarging one's contribution with the uncredited contribution of others. Intentional plagiarism is avoiding to give credit where credit is due. Sir Isaac Newton (a contemporary of Pascal) wrote: *"If I have seen further it is only by standing on the shoulders of Giants"*. Newton and Pascal recognized that much of what they achieved is due to the people who came before them. Pascal was critical of prideful people who claimed to have accomplished everything on their own.

*"Certain authors, speaking of their works, say: '**My** book,' '**My** commentary,' '**My** story,' etc. They are just like middle-class people who have a house of their own on main street and never miss an opportunity to mention it. It would be better for these authors to say: '**Our** book,' '**Our** commentary,' '**Our** story,' etc., given that frequently in these, more belong to other people than to them." (Pascal, Thoughts)*

Plagiarism covers more than plain cut and paste. It includes word substitution. In this case, the plagiarism is definitely intentional. The writer knows the sentence is from someone else. To evade suspicion of plagiarism, the writer changes a word here and there in the plagiarized sentence. Literature has a term for this bad practice: "patchwork plagiarism".

Intentional plagiarism also includes the complete rewrite of a sequence of ideas. The writer reads the original text (possibly written in a foreign language) and rewrites it sentence by sentence using different words. This is wrong. What the law protects is more than the expression of ideas (copyright), it is also their sequence. If I translated a paragraph from a French book, all words would be different, but I would still be plagiarizing. The ideas expressed in successive sentences would be exactly the same.

Even subtler is plagiarism of oneself (self-plagiarism). You might think that it is unnecessary to quote a sentence or paragraph from one of your earlier publications. But, in all likelihood, you assigned your copyright to the journal, in which case the reproduction rights of your article no longer belong to you. Copying large chunks of your past publications (including visuals) would constitute a breach of copyright unless it is authorized. Good news is, all publishers have someone dealing with permissions. So if you want to reuse a visual or table from another paper, especially if it has been published in a different journal, ask for permission. If you are the author, permission is rarely refused, but the publisher may impose some restrictions such as obligatory mentions, etc.

Copyright may still be yours if you published your paper in an open access peer-reviewed journal. To retain this right, writers pay a publication fee per article published. However, the use of open access entails attribution. Open Access journals such as PLoS adopt the Creative Commons Attribution License. This license[1] allows people to download, reuse, reprint, modify, copy, and distribute, *as long as original author and source are mentioned.*

Plagiarism has become such a problem that most journals are now using plagiarism-detection software (Turnitin, Copyscape, Ithenticate) to identify scientists who plagiarize. Research institutes or journals which value their reputation often require researchers to check their paper against plagiarism before submission. It is only a matter of time before such checks are conducted retroactively. Woe to the researchers found plagiarizing, even twenty years ago.

Plagiarism is just one of the crimes of unethical writing. For more crimes such as selective reporting, ghost authorship, and questionable citation practices, do read the document[2] written by Dr. Miguel Roig

---

[1] https://creativecommons.org/licenses/by/4.0/
[2] https://ori.hhs.gov/sites/default/files/plagiarism.pdf

and sponsored by the Office of Research Integrity (ORI). It is entitled: *"Avoiding plagiarism, self-plagiarism, and other questionable writing practices: A guide to Ethical Writing"*.

Now that you are aware of the dangers of plagiarism, let me encourage you to quote. Quoting is good practice. It demonstrates your intellectual honesty. The list of advantages of quoting does not end here. Quotes are proof that you have read the whole paper, not skimmed its abstract. You are seen as someone authoritative, someone who does not take shortcuts. When you give credit where credit is due, you have everything to gain and nothing to lose. Quoting scientists who have been published, particularly if they are authoritative, adds credibility to your own work. If you do not share their views, quoting what you object to cannot be disputed. You do not interpret; you quote.

Observe how Professor Feibelman quotes others.

"In apparent support of the half-dissociated overlayer, Pirug, Ritke, and Bonzel's x-ray photoemission spectroscopy (XPS) study of H2O/Ru(0001) **"revealed a state at 531.3 eV binding energy which is close to [that] of adsorbed hydroxyl groups"** *(28)."*[3]

Note how skillfully he quotes from another paper to hint (with the word '*apparent*') that the support is not there at all. Indeed, the next sentence (not shown here) starts with '*However*' to confirm the lack of support.

TRAP 3 — The Trap of References

Incorrect reference

How many references do you list at the end of your paper? 20, 30, 50? Did you read all the corresponding papers? Where do these references come from?

Let us just take one of the references you added to the reference section of your paper. Where did it come from? A paper you wrote? A review article? A citation index? Let's imagine it comes from the list of references you found at the end of a paper you

---

[3] Reprinted excerpt with permission from Peter J. Feibelman. "Partial Dissociation of Water on RU(0001)", SCIENCE Vol 295:99–102 © 2002 AAAS.

read, and let's follow the trail of that reference like a bloodhound. Where did the writer find that reference? In an older paper they wrote, a review article, a citation index, or in the reference section from another paper? Do you get my drift? The reference that you are using is as reliable as the process used to capture the reference by all the writers who used that reference before you. This process may have been manual or electronic. If the reference was extracted from an older printed paper through OCR (optical character recognition), it may contain errors. If the error occurred during manual entry, it is then propagated electronically, unless someone got hold of the original paper, spotted, and corrected the reference error. But why bother to check? Can't we trust what comes to us electronically?

The news story of Philip Ball in the December 12 2002 issue of Nature created quite a stir when he reported the findings of the paper entitled: *"Paper trail reveals references go unread by citing authors,"* by Mikhail Simkin and Vwani Roychowdhury. In the original Simkin paper,[4] entitled *"Read before you cite!"*, the authors claimed that "only about 20% of citers read the original." The percentage may be too low and the conclusion hasty because one cannot logically infer from the copy of an erroneous reference that the writer has not read the paper. But the problem is real.

The point is this: Even though one cannot infer that the writer has not read the paper when an reference error is found, the reviewer may still have that suspicion. If the writer is suspected of taking shortcuts in his references, then the list of references is no longer indicative that the writer is knowledgeable. The overall credibility rating of the writer goes down.

Imprecise reference

But the problem of bad references does not stop there. Often times a reference is wrong not because it is incorrect, but because it is at the wrong place in a sentence. Imagine this:

---

[4] http://arxiv.org/pdf/cond-mat/0212043v1

Paper [6] proves that the boiling temperature of water is *100°* Celsius.

Author John Smith writes the following sentence:

"It has been proven that water boils at 100° Celsius and freezes at 0° Celsius [6]."

Of course, this is wrong. Paper [6] does not even mention the freezing temperature of water. You would perpetuate the error if you wrote the following in your own paper which references John Smith's paper:

"The freezing and boiling temperatures of water are known [6]."

Had John Smith placed the reference at the correct place, the problem would not have occurred.

"It has been proven that water boils at 100° Celsius [6] and freezes at 0° Celsius."

The guideline for proper referencing is given in *Scientific Style and Format: the CSE manual for authors, editors, and publishers*, now in its eighth edition. It is very clear: "A reference immediately follows the phrase to which it is directly relevant rather than appear at the end of long clauses or sentences." Anything else may introduce imprecision, or worse, error in attribution.

Checking the references takes time. Can't we just trust that other people check them first? Can't we just trust the citation tools in Endnote? Well, rubbish in, rubbish out applies in this case also. In short, check all your references and read the original papers you reference.

Unnecessary references

Writers know that supporting a text with too few references makes it less credible, so they make sure to build out their reference lists. One of the easiest places to add references is in the first few sentences in the introduction when the author establishes the problem. It is easy to pack references at the end of a sentence that expresses a common sentiment such as "pollution is a major disruptor of marine environments [1–12]". While all twelve references may be individually relevant, bulk delivery to the reader is not acceptable. This writer-centered approach to unfiltered referencing creates problems for both the reader and the publisher.

Publisher perspective: The reference list at the end of every article is essential, but it quickly grows long, especially in longer articles with 50–80 references! Publishers only have so many pages available per journal issue to publish research, so they would rather publish meaningful text than reference lists. In fact, many journals now reject reference stuffing in their instructions to authors, limiting to 5 or less references per point. Make sure you specifically check the instructions for your targeted journal before publishing!

Reader perspective: The writer may think they are helping the reader by giving them a plethora of documents to choose from, but there truly is such a thing as too much information in today's time-pressed, results-driven world of research. Expecting the reader to chase down all 12 references for complete understanding is unrealistic. At best, an extremely motivated reader may choose to pursue a handful of these references. But which ones will they pick? Will you leave it up to chance, and hope that they choose the best ones, or the ones that are most relevant to your paper? Shouldn't you, as the writer, carefully present your reader with only the best references from that list?

Unbalanced references

So what makes a reference worth mentioning? References typically come in two flavours. The first type adds credibility to introductory statements. It informs the reader that what you have written is backed up by peer reviewed research. The problem is, one could easily fill out a reference list with references of this type. The second type points to a paper that is closer to yours. It likely shares analysis, insights, or methodologies with your own paper. These references are likely to show up in the introduction, but also throughout the rest of your paper, for example in the methodology, results, or discussion sections. These references are far more valuable to the reader. They show how your work and results contrast and compare with the rest of the scientific field.

☀ Make sure that your introduction contains high quality references that are used more than once in your paper.

Plagiarized references

What is a plagiarized reference? Imagine you've just read a recently published paper on a topic similar to your own, and you believe that they've done a great job of setting the stage in their introduction. They clearly establish the problems faced by the field and have sourced out relevant references already. Why go through all the hard work of finding original sources when you could just re-use their references? It's so easy to just control-c and control-v, and voila, much of the hard work of the literature search disappears. On the face of it, this doesn't seem like it should necessarily be bad behaviour. After all, if the original author has done a good job of choosing relevant references, why is there a need to re-invent the wheel? Let's take a look at why it's still best to put it some hard manual work.

This behaviour is akin to undergraduate students who, when faced with writing on a new subject, head to the nearest wikipedia page and use the references listed there as their sources. The worst offenders never even read the referenced material, they trust that the subject matter they are quoting is relevant to their own specific context. That is quite a gamble. As we stated under the previous subheading of *unbalanced references*, every author should choose references that resonate well with their own research. The more selective they have been, the less transferable these references should be.

This said, references from other papers are excellent sources for you to start your own research, and many references may end up being reused after all. But the odds that you will end up re-using the same references in the same order to make the same points are minuscule.

Missing references

Restraining from citing the work of your closest competitor in research is tempting. You may be competing with them for grant funding or publication. Why boost their citation count and potential visibility on the stage of research? However, not citing your competition is a strategy that is likely to backfire spectacularly.

A reviewer who knows the field is familiar with both you and your competitor already. The reviewer may explain the lack of citation in one of two ways. 1) You are purposefully omitting the competition for strategic and borderline unethical reasons, putting your needs above the needs of your readers. 2) You don't know your competitor. You may be free of any unethical conundrums, but you come across as poorly read.

Between these two choices—unethical or incompetent—you can't win. So cite your competitors. They will need to cite you too!

Courtesy references

Harassment in Science

Today, more than usual, Vladimir stared pensively at his evening coffee. This did not escape Ruslana's notice, who delicately asked him if everything was fine at work.

"Perceptive as always, my dear! No, your husband is just fine. It is my friend Pyotr who is in trouble."

"Trouble? What kind of trouble? Nothing too serious, I hope?"

"That depends on your definition of serious. After leaving our institute, Pyotr found employment in a new lab, and the work habits there are quite different. Within just a few days of starting there, Helena — that's his supervisor — walked up to him and more or less demanded that Pyotr always cites her papers in his own. Can you imagine? She said it was an 'unspoken policy' in the lab. What nose!"

""What cheek", dear. You still need to work on your English idioms."

Unrepentant, Vladimir continued. "Eh, regardless, Pyotr is quite lost. He does not wish to offend his new employer, but he doesn't feel it is ethical to cite his boss as a matter of policy. But it gets worse!"

"How so?"

"Apparently one of his junior colleagues approached him and asked to also be cited in Pyotr's paper! He said that in exchange, he would cite Pyotr's paper in his own paper and in this way, everyone in the lab could grow their careers together! You know like "you scratch my head, I scratch your head!"

"It's "You scratch my back, and I'll scratch yours", Vlad. So your friend Pyotr is feeling the pressure from all sides!"

"That's right, Ruslana. You hit it on the cheek!"

"On the nose, Vlad, on the nose."

What a predicament Pyotr finds himself in. Courtesy references are references that you are asked (nicely or not) to include in your paper, regardless of their worth to the reader. It seems pretty clear cut that this behavior is objectionable, and yet it exists. Dealing with a coworker is a little easier than dealing with a boss, so let's start there.

First of all, it isn't always wrong to cite a coworker, even if they ask for it! Many references in the introduction of a paper are interchangeable, so if your coworker's paper could do the job as well as another's, why not support each other! If, however, their paper is not or mildly relevant, the waters grow murkier. You could gently stand your ground, justifying your denial through your reader-focused approach to referencing. Because you are objectively in the right, ideologically and ethically, it is difficult for them to argue back. Be gentle. Just because you are right does not relieve any of the pressure they are feeling. Offer to cite them in the future whenever it does make sense, and allow them to see that you are not unfriendly, but that your hands are tied by your concerns for the readers.

A boss asking you to cite them in all your papers is a much more clear cut case. It is unethical, but it is also much harder to say no. After all, they may have considerable say in whether or not you keep your job. Your first line of defence should be to offer to place them in the acknowledgements. This may mollify their demand for recognition in an ethical way. If however they insist on citations, you have fewer options.

Fortunately, after much thought, we the authors have come up with a solution to this conundrum: play it off as a joke. If you receive the command to do something clearly unethical, pause for a second, then smile and offer a small laugh, accompanied by a variation on the following words: "Ahaha, I see what you're trying to do here, but I wouldn't do that, I'm an ethical researcher. Was that a test? Did I pass?" It doesn't matter how bad your acting skills are or if your boss can see through your bluff. By phrasing your reply this way, the boss now has a choice to make. Do they still demand to have you cite them, even though it would now be clearly exposed as an unethical act? Or do they back down, using the pretense of humor to back out of the situation without losing too much face? Many would choose the easier option, the second one. As you've also shown yourself to be someone that is difficult to manipulate, they are also unlikely to involve you in such schemes again.

TRAP 4 — The Trap of Imprecision

Under the pressure of a conference- or manager-imposed deadline, you may be tempted to prepare the related work section from abstracts, not from the full text of the papers you had no time to read. Abstracts do not contain *all* the results, they do not mention assumptions or limitations, and they do not justify the methods used. As a result, your sentences will resemble this one: *"Many people have been working in this domain [1–10], and others have recently improved what their predecessors did [11–17]."* Reviewers will see through the smokescreen. Stuffing the reference brackets with reference numbers that exceed three, not only points to abstract skimming, but it may also be symptomatic of insufficient knowledge, shoddy science, or unsound methodology.

Abstract skimming, or dotting your paper with the references of articles you have not read, will hurt you in many ways.

- Errors will creep into your paper.
- Because they find your domain knowledge too superficial, reviewers are tempted to lower the value of your contribution.
- Your research will be disconnected from other research efforts.
- Your story will lack detail, and therefore, interest.
- Readers are usually quick to detect authors who write with authority from the level of detail and precision in their paper. If your words lack precision and assurance, readers and reviewers will doubt your expertise, and question your credibility.

If any of the words from table ☛5 are found in your introduction, you may have fallen into the trap of imprecision. But if these words are immediately justified (*'Several technologies, such as...'*), they are fine.

 Read your introduction, and circle the words you find in the list of imprecise words and other words which you feel are imprecise. Do you need them? How authoritative are you? Can you delete them, or replace them with more specific words or numbers to increase precision? Have you checked your references by going back to their source?

Typically	A number of	Several	Many	Most
Generally	The majority of	Less	Others	A few
Commonly	Substantial	Various	More	Usually
Can & may	Probably	Frequent	Often	...

Figure ☛5
Words that are potential indicators of a lack of precision in scientific writing.

Hedge Words

Of Siths and Scientists

"I can't believe it! Delayed once again!" Vladimir Toldoff was not having a good day. The paper he had submitted for publication had come back with mostly positive comments, but all three of his reviewers highlighted the same point. "All of the reviewers commented on my main claim, saying that it is too early for me to say that my proposed theorem is correct. And yet I am so sure that I am right!"

His supervisor, Popov, was walking by and heard him vent his frustrations.

"Aren't you a big fan of Star Wars, Vladimir?" Popov asked slyly, eyeing the stormtrooper figurine on his subordinate's desk.

"Yes, why do you ask?" responded Vladimir, thoroughly confounded by this sudden twist in the conversation.

"Isn't one of your favourite quotes from that series that 'only a Sith deals in absolutes'? Are you a Sith, or a scientist? No experiment, no matter how conclusive, is enough to establish a scientific theory. It must be confirmed and reconfirmed by others to be trusted as fact."

Admonished but defiant, Vladimir tried to push his luck a little further: "But surely any reviewer looking at these results would agree that this data shows that I am correct!"

"No, Vladimir, the data strongly suggests that you are right. It does not PROVE you are. Had you written "these results strongly suggest" instead of "these results show", none of the reviewers would have disagreed with you and you would be on your way to publication! You still have a lot to learn! Come to my office later. I have just the book for you. It's on hedging".

And with that, Popov walked off. Vladimir turned to his coworker and quietly asked: "Surely, he is not asking me to read a book with all the work already on my plate?"

Without missing a beat, his co-worker responded: "It was perhaps, possibly strongly suggested."

One clear difference we have observed between junior and senior scientists is the ability to masterfully present data while avoiding certainties. Junior scientists often generalize their findings, certain they hold true in all circumstances. Senior scientists, on the other hand, are more keenly aware of the constantly evolving nature of science, and know that what is considered indisputable today may no longer be so tomorrow.

To avoid expressing certainties, writers turn to adverbs such as "possibly" and "presumably", and verbs such as "indicate" or "suggest". These types of words are called *hedge words*, and their judicious use will help you avoid the reviewer's objections.

There are four reasons you will want to use hedges in scientific writing:

1. To express a confidence level.
   The material unexpectedly broke when we raised the temperature to 200˚C. This is <u>likely</u> due to...
2. To open the mind to accept a possibility
   The material unexpectedly broke when we raised the temperature to 200°C. We <u>postulate</u> the error came from...
3. To insure against error or over claim
   The material unexpectedly broke when we raised the temperature to 200°C. We <u>believe</u> this is due to...
4. To let the facts speak for themselves
   The material unexpectedly broke when we raised the temperature to 200°C. A further <u>exploration of the parameters suggests</u>...

Nobel prize laureates James Watson and Francis Crick certainly understood the importance of hedge words in their groundbreaking paper on DNA's structure. Here is a sentence from the conclusion of their paper that has been described as one of "science's most famous understatements":

> *"It has not escaped our notice that the <u>specific pairing</u> we have postulated immediately <u>suggests</u> a **possible** copying mechanism for the genetic material."*[5]

---

[5] Watson, J., Crick, F. Molecular Structure of Nucleic Acids: A Structure for Deoxyribose Nucleic Acid. *Nature* 171, 737–738 (1953). https://doi.org/10.1038/171737a0

Take a moment to read the sentence again and compare it to the four types of hedges introduced above. Each de-italicized, underlined, or bolded phrase corresponds to one or more of the hedge types. Identify them.

Amazing, isn't it? In less than 10 words, Watson and Crick managed to hedge in all four possible ways. The specific pairing suggests (hedge type 4), they postulate (hedge type 2), and theirs is a solution they believe in, but the word "possible" expresses that they are not confident enough to declare it as fact (hedge types 1 and 3). Is it necessary to hedge so thoroughly on all of your research? No. In fact, hedging unnecessarily can lead to a lack of credibility as it makes the author sound unsure. But in this case, these scientists were proposing something so monumentally important that it was better for them to suggest and be proven right than to boldly claim and be proven wrong. They needed to secure the approval of other scientists. As David Locke writes in his book *Science as Writing*, "the new sociologists [of Science] argue that scientific 'knowledge' is knowledge not because it correctly relates the true state of the natural world but because it has been accepted as knowledge by the working body of scientists involved."

TRAP 5 — The Trap of Judgmental Words

Some adjectives, verbs, and adverbs are dangerous when used in the related work section of your paper. The danger comes from their use in judgmental comparisons. Adjectives such as *'poor,' 'not well,' 'slow,' 'faster,' 'not reliable,' 'primitive,' 'naive,'* or *'limited'* can do much damage. Verbs like *'fail to,' 'ignore,'* or *'suffers from'* are just as judgmental. Negations like *'may not'* or *'potentially unable to'* cast doubt without a shred of evidence. They make your work look good at the expense of the work of others who came before you. Sir Isaac Newton did not write: *"If I have seen further, it is because they were all as blind as bats"*. The people whose work you judge will one day read what you wrote about them, and be understandably upset.

Does it mean that all adjectives are bad? No, they are just dangerous. Every adjective is a claim; and in science, claims have to be justified. How would you explain and justify the adjective *'poor'*?

What adjectives (if any) are to be used? Adjectives that compliment (with reason) the authors or their work, adjectives that reflect undisputed public knowledge, adjectives supported by data, visuals, or quotes, and adjectives you define.

Here are eight ways to avoid direct judgment on the findings of a particular paper:

- **State agreement or disagreement** of your results/conclusions with another paper's results/conclusions, or state that your results/conclusions are coherent with, in accord with, or at variance with those found in another paper.
- **Use facts and numbers** to justify your claims. Do make sure to be fair and compare apples with apples.
- **Define your uniqueness**, your difference (nothing is comparable to what you do — maybe because you are exploring an alternative never tried before).
- **Quote another peer-reviewed paper** that independently supports your views (a review paper maybe), or quote the limitations as stated by the authors of the paper you are comparing yours with.
- Show you **improve or extend someone's work**, not destroy it.
- **Bring balance to your views**: recognize the value of an existing method in a sentence's main clause while mentioning its limitation in a subordinate clause as in '*Although this method is no longer used, it helped kick-start the work in this field*'. Avoid inverting the clauses as in '*Although this method helped kick-start the work in this field, it is no longer used.*'
- **Compare visually**, thus avoiding the use of judgmental words. Be respectful of people's earlier work.
- **Change the point of view** and the evaluation criteria. Show that with the new criteria (that you have justified), the method you are comparing yours with is no longer as effective.

Scientists of old were very gracious. Let's learn from Pascal, Benjamin Franklin, and Santiago Ramón y Cajal.

Pascal

Blaise Pascal is not only a great scientist but also a great Christian philosopher, and a man with the right attitude. Here is a translation of one of his thoughts dealing with correcting people's mistakes, followed by a similar recommendation from Benjamin Franklin:

> "When one wishes to correct to one's advantage, and reveal how mistaken someone is, one must observe from which angle that person is looking at things, because, usually, from that angle, things look right, and openly admit this truth, but present the other angle from which the same things now look wrong. The one who is corrected is satisfied for no mistake was made, it was simply a matter of not being aware of other perspectives."

Benjamin Franklin

Extract from chapter 8 of Benjamin Franklin's autobiography:

> "I made it a rule to forbear all direct contradiction to the sentiments of others, and all positive assertion of my own. I even forbid myself, agreeably to the old laws of our Junto, the use of every word or expression in the language that imported a fixed opinion, such as certainly, undoubtedly, etc., and I adopted, instead of them, I conceive, I apprehend, or I imagine a thing to be so or so; or it so appears to me at present. When another asserted something that I thought an error, I denied myself the pleasure of contradicting him abruptly, and of showing immediately some absurdity in his proposition; and in answering I began by observing that in certain cases or circumstances his opinion would be right, but in the present case there appeared or seemed to me some difference, etc. I soon found the advantage of this change in my manner; the conversations I engaged in went on more pleasantly. The modest way in which I proposed my opinions procured them a readier reception and less contradiction; I had less mortification when I was found to be in the wrong, and I more easily prevailed with others to give up their mistakes and join with me when I happened to be in the right."

Santiago Ramón y Cajal

In his book «*Reglas y Consejos sobre Investigación Científica: Los tonicós de la voluntad*», Santiago Ramón y Cajal recommends indulgence because methodology is the source of many errors. He never doubts that the author has talent, commenting that, if the author had had access to the same equipment he used, he or she would have arrived at the same conclusion. In any case, the author's work was published, and his own efforts contributed to the advancement of science, whether they were crowned with success or not.

Read your introduction, and underline the adjectives, verbs, or adverbs you find a little too judgmental or gratuitous. Replace them using one of the eight recommended techniques.

The Deadly Outcome of the Sum of All Traps: Disbelief

Stephen D. Senturia, MIT professor and Senior editor of JMEMS wrote in the June 2003 edition of the journal an excellent article entitled: *"How to Avoid the Reviewer's Axe: One Editor's View"*. He writes: *"A paper is written in order of decreasing believability"*. Therefore, everything in the introduction has to be believable. If the reviewer doubts the origin or sincerity of your words, the accuracy of your sources and numbers, the validity of your claims, the extent of your knowledge, or your fairness in character, then disbelief sets in. It is the fly in the ointment, the tipping point that moves the reviewer's first impression of your paper from neutral to negative.

Why would reviewers trust results or the interpretation of results if they can't even believe introductory statements! Nothing written in the introduction should be perceived as deliberately partial (outdated or omitted reference to research papers from rival groups). And nothing should be perceived as speculative. Naturally, we all know the words that express speculation: *'possibly,' 'likely,' 'probably,'* etc; But Professor Senturia adds a few unexpected words to that list: *'obviously,' 'undoubtedly,'* and *'certainly,'* and other strong words of assurance behind which hide mere speculations.

The reader scientist is critical and suspicious for good reasons. Research is expensive and time-consuming. Before taking on board new ideas from other people's papers, scientists want to be sure that these ideas will answer their needs. All they have to get that assurance are the words of the writer, the peer-review process, and their own experience. Drawing from their prior knowledge, they examine the writer's work, and since they cannot verify everything presented in the paper, at some point, they need to decide whether to trust the writer or not.

In this decision process, you realize that the reviewer of the paper plays a critical role. Good reviewers have developed a sixth sense after reviewing many papers. They know that some writers desperate to be published lie by omission (that is why the judge asks the witness to swear to tell the **whole** truth). Such writers omit mentioning the reference to a paper too close to theirs for comfort. They omit mentioning the known (and often crippling) limitations of their method or results. They omit data. They omit to show the results that do not support their hypothesis. Some of these omissions will only be known after readers discover them while trying to reproduce the research results.

The introduction is a good place for reviewers to deploy their antennas in order to pick up any signal pointing to a lack of knowledge or a lack of intellectual honesty. I remember reading an article on presentation skills than claimed that if only one side of an issue is presented, then believability is in the low 10 %; but if both sides are presented, believability is in the high 50 %. The title of the slide was "fairness". In Science, it would have been "intellectual honesty".

> #### The drug info sheet
>
> *To be really scared, don't watch a horror movie. Instead, go into your medicine cabinet, and read the piece of paper folded in eight sandwiched between the two strips of aluminum holding the precious pills that may cure your headache. Take the time to read the microscopic text to build up some really unhealthy anxiety. The warnings are so overwhelming that if the pills don't cure you, they might just as effectively lead you straight to the emergency room.*
>
> *If the pharmaceutical companies disclose these limitations, it is to avoid lawsuits and to help doctors prescribe the right medicine. Not stating limitations in your scientific paper won't kill anyone, but it might damage your credibility, and your chances of getting published!*

Purpose and Qualities of Introductions

Purpose of the introduction for the reader

- It brings the reader up to speed and reduces the initial knowledge gap.
- It poses the problem, the proposed solution, and the scope in clear terms.

- It answers the why questions raised by the title and the abstract.

Purpose of the introduction for the writer

- It gives the writer a chance to loosen the tie, unbutton the collar, and write in a personal way to the reader.
- It enhances the reader's motivation to find out more by reading the rest of the paper.
- It features the writer's expertise in communication skills, scientific skills, and social skills.
- It enables the writer to strengthen the contribution.

Qualities of an introduction

MINDFUL. The author makes a real effort to assess and bridge the knowledge gap. He respects other people's work and is not judgmental.

STORY-LIKE. Its plot answers all the "why" questions of the reader. It uses the active voice and includes the writer ('we').

AUTHORITATIVE. References are accurate, comparisons are factual, related works are closely related, and imprecise words are absent.

COMPLETE. All "whys" have their "because". The key references are mentioned.

CONCISE. No considerable, vacuous starts, no table-of-content paragraphs, no excessive details.

Introduction Q&A

Q: In their introduction, some writers present their main results, others just present their main goal. What is the best way?

A: Follow the journal guidelines on how to write the introduction. Some ask the writer to present the main results and some recommend not to state the results in order to keep the introduction short, to avoid repetition, and to focus on the goals. So there are indeed two ways to write an introduction, and people have strong preferences for or against each way. Each side has convincing supportive arguments. Here are those proposed by the side supporting repetition of results.

(1) The most frequent argument hinges around the famous quote: *"Tell them what you're going to tell them. Tell them. Then tell them what you told them."* This argument may apply to an easily distracted audience, but can we assume that readers are equally distracted?

(2) Another argument, this one more convincing, is that some journals no longer require a conclusion. The last section of any paper is the discussion. In that case, it may be worth repeating the main result in the introduction.

(3) Some say that many readers only read the introduction of the paper anyway, and therefore you had better mention your results there too... just in case. While this may be true, it is doubtful that readers with this behavior would skip the conclusions (if available).

Amongst those who suggest an alternative to the repetition of the results, Michael Alley advocates *"mapping the document in the introduction"*. He gives the example of a journal article where the author successfully manages to present in story form an overview of the methodology, thereby answering the "Why this way" question. The story reveals the problem and the method used to solve it. It remains silent on the results, but mentions the impact of being able to solve the problem.

My personal view is that the introduction should keep things moving and interesting by mentioning the *expected* results and their *foreseen* impact. If you do have to mention the *actual* results, do so without much detail, and make them part of the *why* story. In all cases, whether you mention the results specifically or not, end your introduction with the main expected outcome of your work.

Q: Can I cut and paste sentences from introduction to abstract?

A: The reader and the reviewer easily detect such shortcuts. Do not give readers the impression that you are in a hurry. The abstract is not written in the same way as the introduction, and the introduction is not written in the same way as the body of your paper. The verb tenses are different, the style is different, and the role of each part is different. Cutting and pasting extends beyond the mere transport of words. It carries writing styles, precision levels, and verb tenses that are fine in their original settings, but not necessarily fine in their new settings.

Q: Can I paraphrase sentences from the introduction of some earlier papers since the context of my new paper is the same?

A: Doing so contributes to boring papers, and you may be found guilty of self-plagiarism. The reason introductions feel repetitive is often because the writer rewrites a paper for different journals, or because there is not much of a knowledge gap filled between two successive papers from the same writer. To avoid such problems, think of each paper as a unique piece of communication to a unique reader. Do not rewrite, write afresh.

Q: When does one write the introduction of a paper?

A: In his excellent little book *"A Ph.D is not enough"*, Professor Feibelman gives sound advice:

> *"Virtually everyone finds that writing the introduction to a paper is the most difficult task. [...] My solution to this problem is to start thinking about the first paragraph of an article when I begin a project rather than when I complete it."*[6]

When you write the introduction of a paper early in your project, you still have the excitement of the journey that lies ahead to energize your writing: the tantalizing hypothesis, the supportive preliminary data, and the fruitful methods. Yet, some argue that the introduction should be written at the end of the paper, once the contribution is clearer. So when does one write the introduction?

Let the content of your introduction dictate the timing. If you write a teaser introduction that states the goals, the context of your work, but

---

[6] Copyright 1993 by Peter J. Feibelman, "A PH.D is not enough: a guide to survival in Science" published by Basic books.

only the expected results and impacts, you can indeed write your introduction early, assuming that your work keeps its original focus. Writing the introduction can also be done early if you prefer to write a succession of short papers, because your limited focus is unlikely to change.

But if your paper is the result of collaboration between many researchers spanning over a few years, it may be impossible to write the introduction early. You would then have to rely on your excellent writing skills to recapture the essence of your earlier goals and motivations to keep the reader interested along a good storyline.

Q: How long is the introduction of a paper?

A: A director of research I knew used to systematically reject any paper in which the introductory segments made less than 30% of the whole paper. To him, these introductory segments are necessary to bridge the reader's knowledge gap. They include the introduction, but also the technical background section that immediately follows it. In the writing seminars we conduct, we rarely see papers with such a thorough background. The majority of papers we see have introductions that represent 10 to 15% of the paper. Occasionally, the sum of introductory segments is slightly above 20%, and for short letters, the introduction is only one paragraph.

So, how long is the introduction? To answer this question without thinking of the reader is unreasonable. The introduction is for the non-expert reader. The writer must have a precise idea of the type of non-expert reader likely to benefit from his or her paper. How much does that scientist know? How much background does that scientist require to use the contribution in whole or in parts? One can make assumptions of the non-expert reader based on journal and keyword choice. Does the title of the paper contain many highly specific terms? Or is your chosen journal very niche, publishing only incremental results in a small field? If so, the interested reader is more likely to be an expert and requires less introduction. If on the other hand you aim to be published in a journal like Science that has a very broad readership, much more introductory material is needed. Note that a paper with multiple contributions is likely to sacrifice the technical background to fit within a given number of pages. That is why it is better to write multiple papers, each with a single contribution and with an appropriate gap-reducing introduction.

Q: What can one tell about a scientist from reading the introduction of their latest paper?

A: Amazingly, one learns a great deal. If the introduction is well developed, and yet easy and interesting to read, the writer displays good communication skills. If the introduction contains targeted references (as opposed to bulk references) and few or no imprecise words, the writer displays good scientific skills. If the introduction contains no judgmental words, the writer displays good social skills. Communication, scientific, and social skills are essential qualities. Besides reviewing the CV or publication record, managers would be well advised to read the introduction from the latest paper written by their potential hire.

Q: How can I give a hint to the reader that my results are not able to reproduce the results claimed in some of the related works?

A: Use the past tense when referring to possibly erroneous findings. Use the present tense to show that, as far as you are concerned, you are in no doubt that the information is correct as shown in this sentence: *Tom et al. identified a catalyst that increases the yield at high temperatures [7].*

What sentence comes next, (1) or (2)?

(1) Slinger et al. subsequently reported that the increased yield is not due to the catalyst [8].
(2) Slinger et al. subsequently reported that the increased yield was not due to the catalyst [8].
The correct answer is (2). It allows you to contradict Slinger's findings with the following sentence:
We found evidence that the catalyst does increase the yield.

Let's change the first sentence by expressing doubt on the findings of Tom *et al.*

Tom et al. identified a catalyst that increased the yield at high temperatures [7].
Slinger et al. subsequently showed that the increased yield is not due to the catalyst [8].
We also found evidence that the yield increase at high temperatures is not linked to the catalyst but to...

Q: In your list of four justifying questions (*why this, why now, why this way,* and *why should the reader care*), I noticed that the question *"why you"* was missing. Do I need to answer that question also?

A: The *why you* question is not answered directly. You have four ways to answer it. (1) Your track record — the trail of references from your previous papers. (2) The acknowledgements — if people fund your research, it is because they are confident that you can deliver something of value. (3) The most senior author mentioned in your list of authors — a well-known and well-cited author acts as a warrant. (4) The reputation of your research institution and its track record in well-cited research papers in your field.

But if you are new to the field, an outlier with no research sponsor, no academic heavyweight on your list of authors, and no glamorous university or research center to attach your name to, do not despair. Even if the odds seem against you, what matters in the end is the quality of your writing and the timeliness and impact of your research.

Do not forget that besides the four reader questions you mention, an additional two sets of questions on problems and solutions come from the reviewer: (1) Is the problem real and is it a useful problem to solve; and (2) Is the solution novel and better than other solutions.

Q: Should I always place the bracket containing a reference at the end of a sentence?

A: The reference must be unambiguous. For example in the following sentence, where should the reference be placed to avoid ambiguity when referring to the review paper, in * or in **?

*Three speech recognition technologies * are prevalent today: Hidden Markov Models, Neural Networks, and statistical methods such as Template Matching or Nearest Neighbor **.*

The correct answer is * because ** would be ambiguous as it may only refer to the nearest neighbor statistical method, not to the review paper that covers all three technologies. In short, the reference should be placed *immediately* after the information it references. If each technology has its own reference, then the following scheme would apply:

Three speech recognition technologies are prevalent today: Hidden Markov Models [1], Neural Networks [2], and statistical methods such as Template Matching [3], or Nearest Neighbor [4].

Q: What references should I put in the reference section?

A: Multiple types of references: (1) recent references because they show that you are keeping up-to-date with what is happening in the field; (2) references to papers published in the journal you are targeting[7]; (3) references to review papers when the information you use can only be found in them, but (4) references to the original paper and not the review paper when the review only points to the original paper without adding value; (5) references to papers YOU HAVE READ; (6) references to all the papers that have directly contributed results, data, or methods to your paper; and (7) references to the papers from the leader in your field.

Q: What references should I NOT put in the reference section?

A: References simply cut and pasted from one paper to the next, references to papers you have not read, references loosely connected to what you are doing, references to mildly relevant papers published by one of the following classes of people: your researcher friends, your manager, people from the same university or institute, or the unknown researchers who cited your paper — just to return the favor.

Q: How do you paraphrase? Is it good to use websites that paraphrase the text you submit?

A: Do not paraphrase while looking at what you want to paraphrase because you could easily introduce plagiarism that way. Instead, just read and get a good idea of what the other paper says. Then, with the paper out of sight, summarize the main points using your own words, and then you add the reference to that paper at the end.

---

[7] Why from the targeted journal? Because you show the editor of that journal that your work is relevant to their interests. But be reasonable — 20–30% of references from the targeted journal is fine. 80–90% might instead give the illusion that you are not well-read and only know their journal.

Unlike humans who benefit from having a deep semantic understanding, paraphrasing websites start from the words in the original text, not from a place of understanding. As a result, many times the paraphrasing introduces errors in your writing, and even sometimes distorts the original facts. Avoid!

Introduction Metrics

✓(+) The introduction starts fast, without warm-ups.

✓(+) The introduction finishes with the anticipated outcome of the research.

✓(+) The introductory segments enable the non-expert reader to benefit from the paper. They represent more than 15% of the paper.

✓(+) References are never in groups exceeding three, the majority are single.

✓(+) All four "why" questions are answered explicitly: why this, why now, why this way, and why should the reader care.

✓(+) The introduction is active, personal, and story-like.

✓(+) Methods, data, and/or application field correctly frame the scope of the paper.

✓(+) The rare imprecise words found in the introduction are qualified immediately after their use ("several... such as")

✓(+) Judgmental words are never used and the story plot is well connected.

✓(+) Background is provided for each specific title keyword.

✓(−) The first sentences of the introduction are known to non-expert readers, or attempt to warm the reader by referring to the hot research topic.

✓(−) The introduction does not finish with the impact of your contribution.

✓(−) Bridging the knowledge gap has not been considered. Introduction size is below 10% for a regular scientific paper.

✓(−) References are presented in groups exceeding three.

✓(−) The answer to one or more "why" questions is missing

✓(−) The mostly passive voice introduction uses less than three personal pronouns.

✓(−) The scope of the paper is mentioned in parts only, and not easily identified.

✓(−) Imprecise words are sprinkled throughout the introduction.

✓(−) Judgmental words are found in the introduction, or the story plot does not include comparisons or does not relate the paper to past papers.

✓(−) Background is missing for some specific and intermediary title keywords.

AND NOW FOR THE BONUS POINTS:

✓(+++) The introduction contains one visual or (+++) the average number of words per sentence in your introduction is 22 words or less.

Chapter 18

Visuals: The Voice of Your Paper

A voice attracts attention; it announces, it warns. It is a substitute to writing: one can read a book or listen to a recorded version of it. Likewise, photos, tables, diagrams, and graphics attract attention even without words. They are worth a thousand words. *The voice gets out of the body. It is not necessary to see the body to hear its voice.* Visuals inform readers independently, even before they start reading the first paragraph. *A voice has its own language, a universal and wordless language, like the one used by the child who babbles, laughs, and cries.* Visuals have their own language, the universal language of scientific graphics. They tell a story directly and quickly with minimum text. *Voice intonation reinforces the message expressed by the body.* Visuals also reinforce the main message of the text, and are in synergy with it.

Just observe the title of this chapter for a few seconds and then bring your eyes back to this mark ☛

Title, headings and subheadings shout, don't they? They are so authoritative in their bold font suit. Framed by white space, nothing crowds them in their spacious surroundings. They are understood at a glance.

Tables and diagrams speak just as much as photos. Guided by a grid of vertical and horizontal lines, bold font, and arrows, the reader captures a large volume of information in little time and easily extracts trends and relationships between the visual elements. The visual story is told in few words.

Visuals excel at comparisons such as *before* versus *after*, or *with* versus *without*. More talents are featured in ☞1

Table ☞1
Typical use for visuals

To represent complexity	To summarize	To reveal sequence
To classify	To reveal patterns	To establish relationships
To compare and contrast	To give precision and detail	To provide context

Readers prefer a visual to text for several reasons:

- In linear text, eyes walk like ants along the narrow path created by lined-up words. In visuals, eyes leap like crickets from one place of interest to another, probing with silent questions. Readers enjoy the speed and the freedom of self-guided exploration.
- Because the text part (title and caption) of a visual complements its graphical part, readers understand more easily.

Visuals have a loud and convincing voice, but only if you can make them speak. Their language is based on a special grammar that describes the correct use of fonts, blocking, kerning, framing, white space, line and color. This visual language is well understood by graphic designers. They can make a visual shout, whereas most of us can only make it whisper or croak. This chapter is not about graphic design, it is about the correct use of visuals in a scientific paper. It is also about principles that will help you design visuals that have an impact from a scientific perspective, even if the lines are a little thin, the white space is not quite well-distributed, or the kerning is an abomination. You may not get an Oscar in a design competition, but you will have visuals that do more than whisper or croak.

However loud and clear a voice may be, if it babbles or utters gibberish, it is not helpful. Visuals need to deliver a loud, clear, intelligible, and convincing message.

Seven Principles for Good Visuals

After reading hundreds of papers, I have detected consistent error patterns. A bad visual breaks one or more of the following principles.

- A visual does not raise unexpected questions.
- A visual is custom-designed to support the contribution of only one paper.
- A visual keeps its complexity in step with readers' understanding.
- A visual is designed based on its contribution, not on its ease of creation.
- A visual has its elements arranged to make its purpose immediately apparent.
- A visual is concise if its clarity declines when a new element is added or removed.
- Besides title and caption, a visual requires no external text support to be understood.

Principle 1: A visual does not raise unexpected questions

As soon as a visual appears, the eyes of the reader probe the visual. They are on a fact finding mission. Wouldn't it be nice to track their path! Using eye-tracking hardware, we've done just that. Four participants were asked to look at a particular visual for five seconds (they did not know what to expect), and then to look at it again, this time in order to ask as many questions as the visual raised.

So, if you do not mind, we would love for you to do the same. Without the eyeball-tracking machine, you will have to remember where your eyes travel in the first five seconds. We are confident that, within that time, your brain will predictably direct your eyes to key parts of the visual; probing, evaluating, and asking silent questions. The first questions will be "what-am-I-looking-at" questions.

It is now time to put your eyes to the test, but do not look for information in the title and caption of the figure because we have removed them to help you focus on the visual itself. Look now at ☛2 for five seconds, blinking your eyes every second, each time memorizing what you were looking at. Afterward, return here and read the next paragraph.

In the first five seconds many things happened. The first one was the discovery of the overall image: an X-Y Diagram with a singular point of change which is meaningless unless you know the meaning and value of its coordinates. Your eyes then moved to one of the two axes.

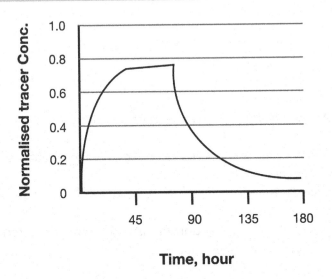

Figure ☞2

Some went to look at the X axis, maybe looked at the end value, then at the unit (hours), and then looked at the Y axis. The others looked first at the Y axis label and tried to decipher the obscure abbreviation, then moved their eyes down to the unit of the X axis (Time), and they moved past the 90 hours to go to the point of change on the curve, and then looked at the Y value for that point. Interestingly, not all went to the Y axis first, even though it is supposed to contain the dependent variable. Those who looked at the X axis first were not driven by logic but by the ease of reading. Horizontal information is easier to read than vertical information.

Look at the visual again, this time without a time limit and identify all the questions this visual raises. Before you look again at ☞2, let me answer the first two questions: What is the Conc. Abbreviation, and why is time so slow? The graph represents the evolution in concentration of a fluorescent tracer in a large tank containing some muddy toxic sludge. This explains the use of hours. Now it is your turn. What other questions do you have? When you are done with your questions, read the next paragraph.

This curve ☛3 looks like a quadratic curve but it isn't. The differences need to be explained.

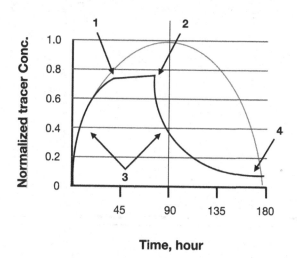

Time, hour

Figure ☛3
I have superposed a regular parabola on top of figure 2. This graph 'asks' five questions. Each arrow corresponds to one question. Can you guess what the fifth question is?

- Question1: Why is the top of the curve linear between points 1 & 2?
- Question 2: What happens at point 2 to change the behavior of the phenomenon so drastically?
- Question 3: Why is the curve convex on the way up and concave on the way down?
- Question 4: Why is the curve asymptotic for high values of time?
- Question 5: What is the concentration normalized to, and why doesn't the curve reach 1?

The readers will look for answers to these questions in the caption. If they are left unanswered, the readers will be frustrated **because the visual raises more questions than the writer is willing to answer.**

Chief among the visuals that raise unneeded questions is the screenshot. It is widely used to illustrate papers because a mouse click is all it takes to capture it ☛4.

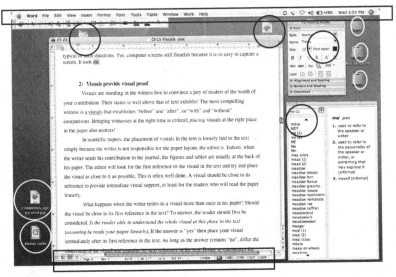

Figure ☛4

Visual gallery of errors exhibit. The cluttered screen 'dump'. If my objective is to show the content of the center window, what are the extra windows, circled files and folders and framed menus doing on this visual?

The screen capture includes all the artifacts of the software application: menu items, windows, icons, tool palettes, and other distracting elements that raise questions.

Naturally, to the writer who captured the screen, everything on the screen is clear and familiar. But is it also clear to the non-expert reader that you have in mind, the one who stands to benefit from your work? There are only two ways to know: ask a reader, or look at each visual of your own paper **as if for the first time** and pretend to be that reader.

Readers come up with the most unexpected questions! What is at their source?

You will often find that they are caused by the reader's lack of knowledge. You see, the writer knows too much! That is why it is useful to ask someone else to do this exercise. You may be able to identify by yourself other sources of questions such as unfamiliar acronyms and abbreviations for example. Once you have identified all questions the reader might have (or has), you have five choices:

- **Leave the visual as it is** but answer the unanswered reader questions in the visual title, its caption, or in the body of the paper.

- **Add to the visual**. Add clarifying visual elements (boxes, arrows, links...)
- **Take away from the visual.** Remove whatever raises distracting questions to keep the focus on the essential point made by the visual.
- **Divide the visual**. Split it into two or more independent visuals that are less complex.
- **Modify the visual** elements (shape, size, order, font, etc) to clearly reveal the purpose of the visual.

The only choice **you do not have,** is to ignore these questions.

What are the questions raised by your visuals? Choose the key visual in your paper (the one which is the most representative of your contribution), and show it to one or two colleagues. HIDE the caption; show only the visual and its title. Ask them which questions the visual raises. As they tell you, write these questions down. When they have no more questions, DO NOT answer the questions. Uncover the hidden caption. Ask if their questions are answered in the caption or in the part of the paper referring to the visual. Once you have identified the problems, ask the reader whether the caption contains new information not illustrated (apart from the description of the visual's context or the visual's interpretation). If the caption does contain visually unsupported information, remove it; otherwise, it would be a claim unsupported by visual evidence.

Show your modified visual to new colleagues and verify that the caption now answers ALL questions, and does not itself raise unwanted questions.

Principle 2: A visual is custom-designed to support the contribution of only one paper

Do you remember this particular visual you worked on for hours? It was a work of art. Using Photoshop, you or the graphic artist had spent much time to make it look perfect. It had been admired at an internal technical presentation or at a previous conference. Part of it does illustrate a key point in your paper, but that part cannot easily be extracted from your original masterpiece without serious rework. You are tempted to reuse the whole drawing/diagram as it is. As a result, the visual includes information irrelevant to your purpose: names, curves, numbers, or acronyms foreign to the reader. All raise questions. Hence, the second principle: a visual is custom-designed to support the contribution of only one paper.

Redrawing is a small price to pay for a custom-designed visual in view of the benefits, not just to the readers, but also to you because (1) your contribution, unclouded by irrelevant details, is easier to identify; and (2) you do not have to ask for the permission to reuse the original if the reproduction rights belong to a journal.

Principle 3: A visual keeps its complexity in step with readers' understanding

Visuals are your star witnesses. They stand in the witness-box to convince a jury of readers of the worth of your contribution. Their placement in your paper is as critical as the timing lawyers choose to bring in their key witness. To know where to place a visual in a paper, take into account the level of understanding of your reader. Complex visuals are more logically placed towards the end of the paper when the reader has filled the knowledge gap. Simpler visuals or self-contained visuals can be placed anywhere.

What happens when you refer to a visual more than once? Assuming readers read your paper linearly from introduction to conclusion, the first time they are asked to look at your visual, they may find it too complex because they have yet to acquire the knowledge that will make the visual totally understandable. A single visual explained here and there in the paper breaks principle #1. It will always raise more questions than the writer is willing to immediately answer.

So ask yourself, why is it necessary to refer to the visual more than once? Is it because you are making multiple points in one large or complex visual? If this is so, divide the one complex visual into several parts (a), (b), (c) to reduce its complexity. Next, make sure readers have enough information to understand *everything* in Figure 1(a) when reaching Figure 1(a) in the text, and do likewise for parts (b) and (c). But if after dividing the visual into parts you realize there is no value in having visuals 1(a), 1(b), and 1(c) next to one another, say for comparative reasons, then divide the visual into separate visuals and have them appear in a just-in-time fashion next to their respective references.

When you send your paper to a journal, its figures and tables are usually at the back of the paper after the references (unless you submit the paper as a PDF file). The person in charge of the page layout will look for the first reference to the visual in your text, and try to place the visual as close to its reference as possible. This is often well done. But a large visual creates page layout problems. For this reason, if you want your visual to be properly placed, 1) design it so that its width matches an integer number of the journal column width — use a template if provided by the journal; and 2) avoid using small serif fonts (Times for example) that cannot be made smaller without a significant reduction in readability, thus limiting the range of options for the placement of your visual. Use fonts without serifs (also called sans serif fonts) like Helvetica, or Arial because of their uniform thickness. When shrunk, their thicker lines do not disappear as fast as the thin lines always present in serif fonts.

Principle 4: A visual is designed based on its contribution, not on its ease of creation

Information with visual impact requires creativity, graphic skill, and time. Because most of these are in short supply, software and hardware producers have provided skill-enhancing, and time-saving tools: statistical packages that crank out tables, graphs, and cheesy charts in a few mouse clicks; digital cameras that, in one click, capture poorly lit photos of experimental setups replete with noodle wires (I suppose the more awful they look, the more authentic they are); and screen capture programs that effortlessly lasso and shrink your screen to make it fit in your paper. The ease of creation of visuals contributes to their abundance—mouse-produced becomes mass-produced. When I show you the photo of a keyboard with the caption "keyboard on which this book was typed," as in ☛5, does it contribute greatly to the usefulness of this book?

Such irrelevant photos are found in scientific papers. They are not useful. They only prove that the writer conducted an experiment using real equipment. To ensure that each visual is critical to your paper, ask yourself whether it replaces much text or strongly supports your contribution. Conciseness applies to both text AND visuals.

Figure ☛5
Gallery of errors exhibit. A Qwerty keyboard, but who wants to know. This photo speaks volumes, doesn't it? It tells you that I used a Macintosh Powerbook with a titanium casing, that I did not use a French keyboard even though I am French, that my right SHIFT key was broken in two, and finally that I am not much of a photographer! What does this have to do with the book itself? Nothing.

The abundance of visuals is a source of unpleasant side effects, the main one being the reader's inability to identify the key visual in a paper. When I ask researchers to read a paper and designate the one *key* graph/figure/table that represents the contribution of the paper, they often disagree with one another and even disagree with the author of the paper. Surely, it should not be so! All should agree on which visual is the most important. Why is there disagreement? It may be due to the author's inability to make his contribution clear visually, but the cause may also lie elsewhere. The more visuals you have, the more your contribution is hard to find, and the more difficult it is for the reader to grasp your whole contribution succinctly. This unpleasant side effect hides another. Removing a visual will weaken your contribution if your contribution is dispersed (diluted) across many visuals.

To summarize, if weak (diluted) visual circumstantial evidence dominates your paper at the expense of succinct but detailed convicting evidence, your article loses clarity and conciseness.

 How many visuals do you have in your paper? Could you identify the one that encapsulates the core of your contribution? Could other people? Are you verbose or concise when it comes to visuals? What do your readers think?

ñ	PostAtl5(liquid)		PostAt30(liquid)		PostAt30(solid)		PostAt30(solid)	
	ß = 1	ß = 20	ß = 1	ß = 20	ß = 1	ß = 20	ß = 1	ß = 20
B4	0.5323+17.4%	0.5323+18.9%	0.4225+19%	0.4254+20.0%	0.2157+9.6%	0.2185+11.1%	0.1493+9.3%	0.1501+8.4%
B6	0.5323+17.4%	0.5373+18.9%	0.4202+18.0%	0.4254+20.1%	0.2156+9.5%	0.2171+10.8%	0.1493+9.0%	0.1500+8.5%
B8	0.5324+17.5%	0.5373+18.9%	0.4189+17.7%	0.4255+20.0%	0.2156+9.5%	0.2182+10.9%	0.1492+9.1%	0.1496+8.5%
BJI	0.4720		0.3706		0.2997		0.2380	

Table 1 Statistics on ß – 1 or 20, LG = 3

ñ	PostAtl5(liquid)		PostAt30(liquid)		PostAt30(solid)		PostAt30(solid)	
	ß = 1	ß = 20	ß = 1	ß = 20	ß = 1	ß = 20	ß = 1	ß = 20
B4	0.5456+23.4%	0.5423+23.9%	0.4324+26.0%	0.4341+27.0%	0.2192+12.6%	0.2232+14.9%	0.1544+13.3%	0.1554+14.8%
B6	0.5440+23.4%	0.5473+23.9%	0.4332+25.0%	0.4341+27.1%	0.2193+12.5%	0.2235+14.1%	0.1544+13.3%	0.1550+14.6%
B8	0.5424+23.5%	0.5473+24.0%	0.4299+24.7%	0.4341+27.0%	0.2126+12.6%	0.2231+14.7%	0.1543+13.0%	0.1556+14.0%
BJI	0.4720		0.3706		0.2997		0.2380	

Table 1 Statistics on ß – 1 or 20, LG = 4

ñ	PostAtl5(liquid)		PostAt30(liquid)		PostAt30(solid)		PostAt30(solid)	
	ß = 1	ß = 20	ß = 1	ß = 20	ß = 1	ß = 20	ß = 1	ß = 20
B4	0.5400+22.1%	0.5323+18.9%	0.4195+27.2%	0.4254+28.5%	0.2198+13.2%	0.2241+15.0%	0.1545+13.8.	0.1580+15.2%)
B6	0.5401+22.5%	0.5349+18.9%	0.4242+28.0%	0.4254+28.5%	0.2198+13.1%	0.2140+15.8%	0.1546+14.0%	0.1550+15.1%
B8	0.5397+22.7%	0.5373+18.9%	0.4201+27.7%	0.4255+28.5%	0.2130+13.2%	0.2235+14.9%	0.1546+13.8%	0.1567+15.7%
BJI	0.4720		0.3706		0.2997		0.2380	

Table 1 Statistics on ß – 1 or 20, LG = 5

Figure 6

Visual gallery of errors exhibit. The visual salvo. A large number of visuals are set side by side; each one is marginally different from the one that precedes, so much so that the eye can barely see the difference between them. In this case the visuals are tables, but they could be graphs or images also.

Principle 5: A visual has its elements arranged to make its purpose immediately apparent

The visual salvo is a popular classic in the gallery of errors leading to complex visuals (☛6).

Visual ☛6 is impressive, but the reader is not impressed. The principle — a visual has its elements arranged to make its purpose immediately apparent — is certainly not applied here.

Clarity of purpose on the writer's side is essential to achieve clarity of meaning on the reader's side. Clouding the meaning is easy: simply bury the key information in the midst of other data so that its purpose does not stand out. For example, writers could inadvertently camouflage key information by presenting the data in the wrong sorting order, by aggregating it in classes that absorb the interesting variance, or by placing it in the midst of loosely related data.

Key data must stand out. They are your crown jewels, but they need a crown; and the role of the crown is not to hide the jewels, but to make them stand out by giving them the appropriate structure. That structure is the arrangement of data that makes clear a purpose. So when visuals fail to reveal their purpose, it is for several reasons:

- The writer has no specific purpose — the visual is a data dump.
- The writer has a purpose, but leaves to the expert the work of identifying that purpose from the data presented. The writer has not considered that the non-expert reader is not able to identify the purpose without his or her help.
- The writer has a clear purpose (it is even mentioned in the caption), but the arrangement of the data does not support the purpose, and the reader struggles to reconnect purpose and data.

To highlight the main point of a visual, its elements need to be organized. In table ☛7, the caption asks the reader to compare one-step with two-steps methods. Even though the arrangement of the table elements is not haphazard with all one-step methods grouped at the end of the table, the reader is lost. What is to be compared? The average value of one-step methods to the average value of two-step methods? Each one-step method with its corresponding two-step method where it reappears as the second step (for example, MO

Table ☞7

Visual gallery of errors. The table without a clear message. Comparison of all combinations of one — two step methods

Methods	True positive rate (%)	False positive rate (%)
BN&BN	22.0	1.3
BN&MO	24.9	1.9
BN&MSV	39.2	0.2
PSY&BN	27.1	2.6
PSY&MO	27.0	2.7
PSY&MSV	66.9	0.3
COR&BN	23.0	1.9
COR&MO	25.8	2.5
COR&MSV	38.1	0.2
BN	21.8	1.2
MO	24.8	1.9
MSV	35.9	0.2

compared to BN&MO, COR&MO, and PSY&MO)? The best one-step method with the best two-step method?

The table, filled with acronyms defined elsewhere in the paper, is unfriendly to the memory. What point did the writer want to make? What do you think? Look again at ☞7, and then return to the next paragraph.

Most readers thought the writer wanted to show that one combination of two steps, PSY and MSV, has the highest true positive and a very low false positive score. If the writer had wanted to make that

Table ☞8

The table's message is clearer. Best method for all combinations of one — two step methods

Methods	True positive rate (%)	False positive rate (%)
BN	21.8	1.2
BN&BN	22.0	1.3
COR&BN	23.0	1.9
MO	24.8	1.9
BN&MO	24.9	1.9
COR&MO	25.8	2.5
PSY&MO	27.0	2.7
PSY&BN	27.1	2.6
MSV	35.9	0.2
COR&MSV	38.1	0.2
BN&MSV	39.2	0.2
PSY&MSV	66.9	0.3

point *immediately* clear, it would have been better to sort the scores in ascending order as in table ☛8.

But if the writer had indeed wanted the reader to compare methods where the one step reappears as a second step, then the following table (☛9) would have been more to the point.

Table ☛9

The table's layout is better to help the reader compare one and two step methods.

Methods	True positive %	False positive %
BN (1-step)	21.8	1.2
BN&BN	22	1.3
COR&BN	23	1.9
PSY&BN	27.1	2.6
MO (1-step)	24.8	1.9
BN&MO	24.9	1.9
COR&MO	25.8	2.5
PSY&MO	27	2.7
MSV (1-step)	35.9	0.2
COR&MSV	38.1	0.2
BN&MSV	39.2	0.2
PSY&MSV	**66.9**	**0.3**

What if the writer was not strictly interested in comparing steps, but in a new point altogether? What if the author was interested in highlighting the most effective second step? The accent is now on *the difference* between one step and two steps. Look at table ☛10.

From table ☛10, the reader immediately sees that the PSY method is systematically the best method as a second step, and that MSV is the best one step method to accompany it. Compared to the original table ☛7, much raw data has disappeared, and in particular

Table ☛10

The table's message is clear and more instructive even though less data is displayed. Comparison of all combinations of one — two step methods

One step method	% True positive rate*	2nd step method (BN, COR, or PSY) Best second step	% Increase in true positive rate (PSY as 2nd step)
BN	21.8	PSY	24.31
MO	24.8	PSY	8.87
MSV	35.9	PSY	86.35

* Maximum false positive rate for all one step / two steps combinations is less than 2.7 %

the false positive rates. The writer has not left them out. Their maximum value is given in a footnote.

Imagine now that each step is a complex clustering algorithm, so that adding a second step requires additional computational resources. The author wants to answer the question "is adding a second step

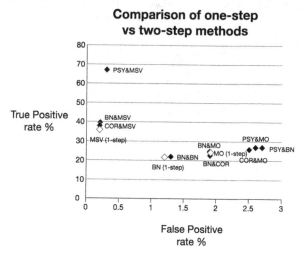

Figure ☛ 11
The comparison of one and two-step methods reveals that the one-step MSV method (36,5% true positives) is superior to the BN and MO methods. MSV also acts in synergy with the PSY method to nearly double the score with 67.5% true positives while keeping false positive rates low.

Figure ☛ 12
The comparison of one and two-step methods reveals that the one-step MSV method (36,5% true positives) is superior to the BN and MO methods. MSV also acts in synergy with the PSY method to nearly double the score with 67.5% true positives while keeping false positive rates low. (One-step method is empty diamond, two-step method is full diamond)

worthwhile?" Two visuals would rapidly make that point (☞11, and ☞12). Note how the caption has been rewritten to help the reader.

In summary, for each point you want to make, find the most appropriate visual. A specific visual makes a specific point; a different point would be made by a different visual. Choose your data based on their added value towards your contribution and based on their conciseness (i.e. they make the same point with less elements). Once chosen, arrange your data until their organization clearly makes your point. This usually takes many drafts. Granted, improvement takes time, but it is time well spent: a visual is so much more convincing than a paragraph.

Principle 6: A visual is concise if its clarity declines when a new element is added or removed

Each visual has an optimum conciseness. Visuals ☞11 and ☞12 include the false-positive values. Are these critical to the point the author wants to make? Is it possible to come to the same conclusion without them as in ☞10?

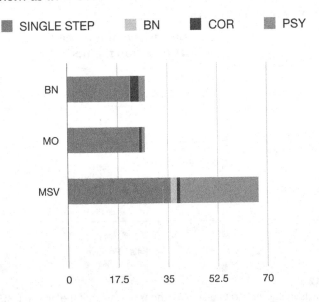

Figure ☞13

Look at ☛13. It contains all of the same information as Figure 11, but is considerably more compact. It makes the same point with three bars instead of twelve. Orienting it horizontally instead of vertically also ensures that it will always fit within one column.

To add visual elements to a graph is tempting; to merge two graphics to make room for more text is irresistible. The resulting visual is so complex that it is no longer understandable. It becomes the "everything but the kitchen sink" visual. The density of its elements per square inch hinders rather than facilitates understanding.

The draft diagram in ☛14 had attractive rainbow colors, 3D elements, arrows, links and much more. Some may recognize the yeast cell growth cycle at its center.

It required much time to design and was a masterpiece. It established a parallel between two phenomena that shared the same cell cycle. It would have been perfect, but for a small problem: only its

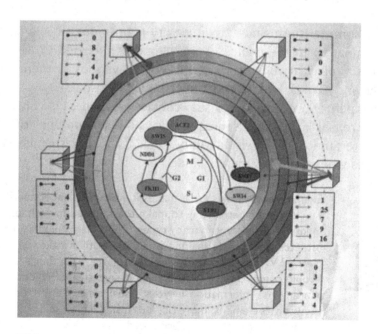

Figure ☛14
Visual gallery of errors. The overly complex visual. This beautiful visual combines two related visuals into one. The resulting increase in complexity greatly reduces clarity and understanding.

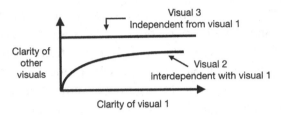

Figure ☞15
If the clarity of a visual is dependent on the existence of another visual, then the two visuals are interdependent, and as a result, the clarity of one visual will affect the clarity of the other.

authors understood it. This figure was later simplified, and clarity returned when visual elements were taken away.

If adding or subtracting elements in one visual affects the clarity of another visual, then these two visuals are clearly interdependent (☞15). They have to be redesigned to increase their independence.

To summarize this point, complexity is born out of (1) a lack of discrimination in the choice of the elements included in a visual, (2) a lack of clear relationship between the various elements in a visual, and (3) a lack of independence between visuals.

Simplification decreases visual conciseness whereas consolidation increases it. But conciseness is always the servant of clarity, not its master; that is why a visual can be considered concise if its clarity declines when a new element is added or removed.

Examine each visual. What makes it difficult to understand?

Is there a better way to make the same point with less elements? Visuals come in many types: chart, diagram, table, photo, list, etc. Would replacing one type with another make your visual clearer? Would dividing one visual into two make things clearer? Would combining two visuals make things clearer? Would reorganizing the information in your visual make the relationships between its elements more obvious (using arrows, colors, words, or sorting the data in a different order)?

Principle 7: Besides caption and title, a visual requires no external text support to be understood

The strange oasis

An old Bedouin likes to tell the tale of a strange oasis he once came across in the Sahara desert where a sand storm had stranded his caravan. The tallest man, who was perched on top of the tallest camel in the caravan, saw it first. "Oasis straight ahead!" he shouted. They pressed ahead. A short distance away from heat and thirst relief, the travelers noticed that clusters of full coconuts were sitting on the sand dunes, away from the coconut trees in the oasis. Their skin was soft, but they were hot to the touch, so the people took them inside the oasis to drink later. The oasis was small, it had no well, and all its coconut trees were barren, so the only refreshment would have to come from the coconuts found on the sand dunes. Unfortunately, these coconuts were not ordinary. Their soft skin hardened like steel as soon as they were inside the oasis in the shade. The sharpest dagger could not cut them open. So people had to go back out into the desert to open them, and return to the oasis to drink them, a process they found most unpleasant.

The oasis is still there, the Bedouin claims. It is now an attraction for tourists who go and visit it by helicopter (camel rides are just too slow nowadays).

In our story, the oasis is the refreshing visual and its caption, the desert is paragraph text, and the coconuts are what should have been in the caption, but was turned into paragraph text instead. Nowadays, readers are pressed for time. So they parachute themselves directly into your paper with a marked preference for the pleasant visuals, which are far more refreshing that paragraph text. However, they are frustrated because, to understand them, 1) they need to refer back to the text and search for the 'see Figure X,'; 2) they then need to locate the beginning of the sentence; and finally 3) they need to go back and forth between the explanatory text, the visual, and its caption, until their understanding is complete. This time-consuming and iterative process is most unpleasant. In the immortal words of the Bedouin, "Coconuts are meant to be inside the oasis". We no longer live in the days of silent movies. A visual must "tell all" by itself, without the need for text outside of its caption. To accommodate the non-linear reading behavior of scientists, each visual should be self-contained, which means self-explanatory. Some will object that this creates redundancy because the text in the paragraph is then repeated in the caption. They assume that the text in the body of the paper remains the same once the caption of the visual makes it self-explanatory. This is not the case. The paragraph in the main section, now shorter, only states the key contribution of the visual without detail, repeating only what is necessary to move the story along. Doing so has two advantages: (1) the visual can be understood without the need to read the whole article and (2) the body of the paper is shorter (and thus faster to read) because it keeps to the essential.

In the original visual in ☛16, the CALB and MCF-C18 acronyms are undefined.

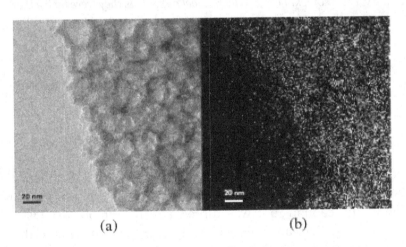

(a)   (b)

Figure ☛16
A visual that does not stand alone and is not self-contained.

(Caption) Figure 5. (a) TEM micrograph of CALB/MCF-C$_{18}$ from pressure-driven method, and (b) the corresponding ELS elemental mapping of N.

Readers have to go back and forth several times between text, visual, and caption before they understand and "get the full picture."

(Text in body of paper)

"Figure 5 illustrates the uniform nitrogen mapping over the CALB/MCF-C18 sample, indicating the homogeneous distribution of the nitrogen-containing enzymes within the mesoporous silica matrix. CALB/MCF-C18 also showed PAFTIR peaks at 1650 cm-1 and 3300 cm-1 (Figure 3c), which were associated with the amide groups of the enzymes, confirming the enzyme incorporation."

Compare the visual with the modified one in ☛17. The new visual is now autonomous, its caption is longer, but the description in the body of the paper is cut down to the bare essentials.

(Modified caption) **Figure 5.** *(a) Transmission electron micrograph of the Candida Antarctica Lipase B enzyme (CALB) immobilized by pressure in the porous matrix of hydrophobic mesocellular siliceous foam (MCF-C$_{18}$), and (b) the corresponding electron energy loss spectroscopy map of nitrogen. Nitrogen is abundant in enzymes. Its detection is used as evidence of their presence. Here, Nitrogen, and therefore the CALB enzyme, is seen as uniformly incorporated and distributed in the porous matrix.*

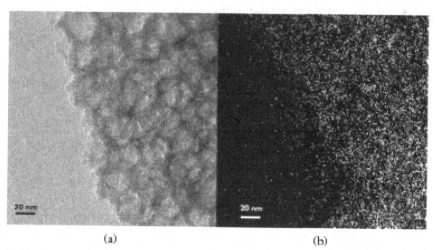

(a)   (b)

Figure ☛17
A visual that is now stand alone and self-contained.

(Modified text in body of paper)

"Both electron energy loss spectroscopy (Figure 5b) and FT-IR spectrum of CALB/MCF-C$_{18}$ (Figure 3c) confirm the incorporation of the enzyme in the Siliceous foam."

The captions of Figure 5 (a) and (b) are now self-contained, while the text in the body of the paper is cut to focus only on the point the writer wishes to make (the enzyme is trapped in the porous matrix). Even if the reader goes straight to Figure 5 and bypasses the text, the same message is given.

Examine each visual in your paper.
Rewrite the caption to make your visual self-contained.
Revise the description of the visual in the body of your paper to shorten it by stating only the purpose of the visual, or its key contributive point.

Purpose and Qualities of Visuals

Purpose of the visual for the reader

- It allows self-discovery of the paper.
- It helps the reader verify the claims of the writer.
- It saves reading time by allowing faster understanding of complex information and faster understanding of problem and solution.
- It provides a direct (shortcut) and pleasurable (memorable) access to the writer's contribution.

Purpose of the visual for the writer

- It makes the paper more concise by replacing many words, particularly in the introduction where it provides fast context, and helps bridge knowledge gaps.
- It motivates readers to read more, yet allows them not to have to read all.
- It provides compelling evidence, in particular evidence of contribution.
- It enables the writer to represent complex relationships succinctly.
- It (re) captures the reader's attention and improves memory recall.

Qualities of a visual

SELF-CONTAINED.
Besides title and caption, no other element is necessary to understand it. It answers all reader questions.
CLEAR and CONVINCING. It has structure, it is readable, and it includes visual cues to help readers focus on key points.
CONCISE. It contains no superfluous detail. It cannot be combined with other visuals without loss of essential information or clarity, nor can it be simplified.
RELEVANT. It has purpose and is essential to the contribution. It does not distract the reader.
QUICK TO GRASP — It should be understood in less than 20 seconds.

Examine each visual in your paper. Is it concise? Can you hide details in the appendices or footnotes? Is the visual essential? Is it rapidly understandable to a reader who is not an expert in your field? Is it autonomous and understandable without any support from your main text? Should it appear earlier or later in the paper?

Visuals Q&A

Q: I would like to recycle a visual I first published in the extended abstract of a conference proceeding to reuse it for a journal publication. The two papers are very close. Can I just reuse it or do I need to redo it?

A: Unless you have kept the copyright to the original paper, you are obligated to ask the owner of the conference proceedings for the permission to re-use the visual. If the two papers are very close, you also have to give the journal a copy of the original conference paper so that the editor can assert whether there are enough differences to merit publication. In short, redo the visual. There are many ways to present data, and I am sure that the original visual could benefit from another draft. After all, any paragraph published, or any visual published is only a draft. There are so many ways to improve on text, likewise for visuals.

Q: Can I change the contrast of my visual to make it more readable?

A: Any image post-treatment is always looked upon with suspicion, and is often forbidden by the journal, so check the journal's instructions to its authors regarding artwork, and abide by them. There is some tolerance, for example to improve contrast, but again, 1) if improving contrast removes information from your visual, or conceals it, you should not do it (increasing the contrast may remove a gray level that was significant); and 2) you should inform the journal that you have manipulated the image, and provide the original image for comparison.

Q: Can a reference to a figure (say figure 2) be made in the caption of another figure (say figure 3)?

A: Preferably not. Figure 3 is not standalone because understanding it requires the prior understanding of figure 2. However if you cannot make figure 3 stand-alone, do mention figure 2 in its caption. It will at least guide the reader to find all relevant information.

Q: Can I write the interpretation of the visual in its caption, or do I leave that to the discussion?

A: If the visual is in the result section of your paper, you are not expected to interpret it there. But remember, the way you have laid out your data already guides the reader towards your interpretation. Highlight the salient data or points that will be used in the discussion section when you offer your interpretation. If the visual is in the discussion section, nothing should stop you from providing an interpretation in the caption. This will definitely help you achieve the worthwhile goal of making any visual stand-alone.

Q: Do I need to bother about page layout issues for my visuals since the journal is going to redraw them anyway?

A: You should if you want to increase the chances for your table or graph to be kept close to its reference in the text. This is particularly relevant when the journal you are targeting has multiple columns on one page. When that is the case, you may want to redesign a horizontal table to make it fit vertically on one column of a two column page. For example, Visual ☛11 in this chapter could easily be made vertical by inverting the x-y-axis with no loss of meaning! But keep vertical what people expect to see vertical, like temperature bars; and keep horizontal what which people expect to see horizontal, like distances. And while on the subject of visuals, never hesitate to insert additional space or lines to add structure and to highlight what you consider important. Use (larger) sans serif type fonts if you expect that the visual will have to be shrunk to fit the size of a page or column. Better still; rework your visual to make it fit on a page or column naturally by increasing conciseness and clarity.

Q: There are so many ways to visualize my data, so which way is the best?

A: Before you even ask yourself that question, ask yourself these, IN THE ORDER GIVEN:

- Which point do I want to make in my visual? Asking this question helps you avoid the pointless data dump.
- What data do I absolutely need to make my point, and which data can I leave out?
- Do I have the data I need to make the point I want to make? If not, what point can I make with the data I have?
- Which level of data aggregation or transformation (frequency, percentage, mean, cumulative, log, etc) will best make the point?
- Now that I have the right data, and I know what point to make with it, what is its dimensionality (2D, 3D, nD vector)? What is its nature (qualitative, discrete, continuous, temporal, pictorial, numerical, symbolic as a chemical notation or the elements in a diagram)? What is its accuracy (error bar, range, probability, resolution)? What is its range and scale (finite, infinite, zero to one, set of known attributes, set of given names, etc)?
- Given my type of data, which ordering scheme would support my point: the data intrinsic order (chronological, numerical, spatial, logical, hierarchical), or a new order such as functional proximity, functional classes, inclusion — exclusion, with — without, before — after, generic — specific, simple — complex,

most probable — least probable, low priority — high priority, most favorable — less favorable, most relevant — less relevant, similar — different, parents — children, etc)?

- Which visual representation is the one the reader expects to see in order to be convinced by my ordered data? (more than one may apply — use the one which makes the point most immediately and most clearly) Table, list, line chart, stacked line chart, flowchart, bar graph, photo, Venn diagram, block diagram, tree structure, schematic diagram, 2D bar chart, 3D line chart, Pie chart, doughnut chart, 2D area chart, 3D floating bars, point chart, bubble chart, grid surface chart, mesh surface chart, radar chart, polar chart, high-low chart, error bar chart, funnel chart, box chart, XYY chart, Western blot, Northern blot, Pareto chart, scatter plot, and so many other representations are available.

Q: Why do I get lost when I look at some visuals, and what should I do so that my readers do not get lost in my visuals?

A: Contrary to the paragraph, which has only one point of entry (its first words), a visual can have many points of entry. Unguided, the reader's eyes dart here and there, trying to find what there is to learn from that visual. You must design the visual such that the eyes are guided in their exploration through the visual. The title of a figure or table serves the same role as the topic sentence in a paragraph. It helps the reader choose the right entry points. The photo of enzyme entrapment in siliceous foam could be titled "Evidence of Enzyme nitrogen (b) trapped by pressure in siliceous foam (a)". Depending on the point you want to make, the title of the one-two step comparison table could either be "The higher the one step positive rate, the higher the rate increase from the second step PSY", or "Adding a second step is not computationally efficient except for the MSV-PSY pairing". While visuals allow for free visual exploration, these titles guide the reader's eyes onto a specific path important to you, the writer.

Q: How do I guide the reader towards the main visual in my paper?

A: Make it stand out. You could make that visual the largest one, or the only one with color. You could put many words from the title of your paper in the visual's title or in in the first line of its caption, or you could simply start the caption with "This figure represents the core of our contribution."

Q: Should I repeat what the X and Y axis mean in the caption?

A: There is no need to repeat whatever is blatantly obvious to all in the figure itself. Write instead on whatever is NOT obvious and is necessary to make that visual understandable and stand-alone. Do not repeat the figure; explain it.

Visuals Metrics (calculate your score for each visual)

✓(+) The visual is self-contained.

✓(+) The visual has no acronyms (in caption, title, or inside visual)

✓(+) The visual is essential to support the contribution and add to its value.

✓(+) The point made by the visual is seen by the reader within 20 seconds.

✓(+) The visual easily fits on one column.

✓(+) The visual's title, table headings, axis legends are clear and informative.

✓(+) The type of visual used corresponds to what the reader expects, or is better than what the reader expects. It fully supports the point made in the text or in the title/caption.

✓(+) The visual does not raise more questions than the writer is willing to answer.

✓(+) The caption provides context to help understand the visual.

✓(−) The visual's understanding depends on external support other than caption/title.

✓(−) The visual has acronyms.

✓(−) The visual is not essential and makes a secondary point.

✓(−) The point made by the visual is missed, or unseen, or slow to see.

✓(−) The visual requires much space and could be far from its reference in the text.

✓(−) The visual's title, table headings, axis legends are cryptic or too abbreviated.

✓(−) The type of visual used does not correspond to what is required to make the point made in the caption.

✓(−) The visual raises more questions than the writer is willing to answer.

✓(−) The caption is silent on the context.

AND NOW FOR THE BONUS POINTS:

✓(+++) Reader and writer, or two independent readers agree on which single visual represents the core of the contribution.

Chapter 19

Conclusions: The Smile of Your Paper

After ruling out many choices, I decided that no part of the body could represent the conclusions better than the smile. Why a smile? I thought again of the many conclusions that had disappointed me, and deflated my enthusiasm with self-deprecatory endings such as 'In order to test real performance improvement....,' or 'This could be greatly improved by...'. I had read these articles with great interest until, in the conclusions section, I had found suggestions that nothing significant had been accomplished. I felt like the person about to buy a car described as safe, only to be told at the last minute that the car had no air bags and no antilock braking system. Previously undisclosed limitations disguised as future work frequently surface in the conclusions to disappoint the reader who genuinely assumed the author had dealt with them already. Imagine a lawyer who managed to demonstrate the innocence of his client throughout the court proceedings, but who, on the very last day in front of the jury, apologizes because not enough evidence has been produced to justify the plea of innocence. How unbelievable!

The way a defense lawyer ends his plea in front of a jury should also be adopted to end a scientific paper: with assurance, firmly, and smiling, trusting the jury will find the client not guilty of scientific insignificance. Lawyers know that the day the jury gathers for the final plea is an important day. The day the writer writes the conclusion of his paper is also an important day. The writer cannot write it at nighttime, close to exhaustion. His writing will be lifeless. The writer cannot write the conclusion too long after the end of the research. His sense of past achievements may be gone.

So, before writing the conclusion, the writer has to re-energize his pen by reading again introduction and discussion to identify the research milestones. He has to accumulate the intermediary scientific merit points to form his final score: the paper's global contribution. Re-energize yourself, smile as you consider your score, stay positively charged... because the negative ions are there: your fatigue, the time that passed since the end of your research, and the limitations that need to be addressed in the future. However, do not waste that positive energy on yourself, to polish your halo, or bask in the sunshine of your glorious past. The conclusion is not an opportunity for an ego trip. It is an opportunity to polish, not your halo, but your diamond of a contribution because you need to sell it to cash in on citations, to encourage others to use your work.

You may have noticed that, in some journals, articles have no conclusions heading; the discussion ends the paper. The need to conclude is still there however, even if the heading is absent. Some journals — Nature is one of them — recommend to finish an article without conclusions. They would rather have the author write the last paragraph "about the implications of what the reader has read",[1] and not summarize what has been accomplished. Professor Railsback with great common sense writes: *"Conclusions are just that — the inferences that can be drawn from your data, not a reprise of the entire paper."* Combining expert advice, 'inferences' and 'implications,' it becomes clear that conclusions are not just another abstract.

Abstract Versus Conclusions

Surely readers wouldn't notice that your abstract is similar to the conclusions, would they? They would!

Readers read in a non-linear fashion. They tend to skip large sections of a paper, jumping from abstract to conclusions for example, like the hurried reporter only attending the first and last day in court. From a writer's perspective, that behavior is not ideal, but the writer can use this knowledge to his or her advantage. First, the writer

---

[1] Nature Physics, "Elements of Style", editorial, Vol. 3 No. 9 September 2007.

now understands how dangerous it is to copy and paste sentences between abstract and conclusions since the reader immediately notices them. Second, the writer should differentiate the conclusions from the abstract to avoid boring the reader. How do the two differ?

- Sometimes, the journal recommends the use of the past tense in the abstract. Unfortunately, the main tense used in the conclusions is also the past tense because you are referring to what you did. Only the facts that have been demonstrated without a doubt, the unquestionable scientific facts, are stated in the present tense. The lawyer says *'my client is innocent,'* not *'my client was innocent'*. The present tense in the conclusions reinforces your contribution. If the journal does not impose the use of the past tense in the abstract, it becomes advantageous to write the whole abstract in the present tense because doing so differentiates conclusions from abstract.

- Whereas the abstract briefly mentions the impact of the contribution, the conclusions dwell on this aspect to energize the reader. In his book *A Ph.D. Is Not Enough*, Professor Feibelman gives his writer's viewpoint.

  > *"The goal of the conclusions section is to leave your reader thinking about how your work affects his own research plans. Good science opens new doors."*[2]

- Conclusions are more comprehensive than the selective abstract. Conclusions bring closure, not on what the self-contained abstract announced, but on what the introduction and discussions opened. They close the door on the past before they open *'new doors'* onto the future.

- The abstract adopts a factual, neutral tone. The conclusions keep the reader in a positive state of mind. Remember that a reader needs strong motivations to read the whole paper, not just your conclusions. The motivating role is traditionally taken by the introduction, but if the reader skips it and jumps directly to the conclusions, then they must also

---

[2] Copyright 1993 by Peter J. Feibelman, "A PH.D is not enough: a guide to survival in Science" published by Basic books.

motivate the reader to reach inside your paper. Therefore, keep your energy level high and think positively about your contribution.

- Everything in an abstract is new to the reader. In the conclusions, nothing is. The conclusions do not surprise the reader who has read the rest of your paper. Following the analogy with the defense lawyer's final plea in front of a jury, it stands to reason that any attempt to convince the jury at the last minute with evidence that has not been cross-examined is not receivable and is objectionable. Such last-minute-theatrical surprises are the realm of Hollywood movies only. Even the section about future work is expected. In the discussion section, you venture explanations that require future validation, or you suggest that different methods might be helpful to bypass constraining limitations. The reader who has read your discussion anticipates that, in your future work, you will explore these new hypotheses or use these different methods.

Examples and Counterexamples

Examples

In the following example, the author repeats a main aspect of his contribution already announced in the discussion section. It is an encouragement for others to use his method.

> Our method has been used to determine the best terminal group for one specific metal-molecule junction. In addition, we have demonstrated that, in principle, it is applicable to other metal-molecule couplings.

It is not always necessary to have conclusive results to conclude. Sometimes, the hypothesis presented in the introduction can only be partly validated. The choice of words to say so is yours, but you must admit that the phrasing is quite critical here. Which of these sentences is better?

> In conclusion, our modified gradient vector flow failed to demonstrate that...
> In conclusion, our modified gradient vector flow has not been able to demonstrate that...

Or

In conclusion, our modified gradient vector flow has not yet provided definitive evidence for or against …

The last sentence is much better, isn't it? The word *"yet"* suggests that this situation may not last. Far for being despondent, the scientist is hopeful. *"Yet"* creates the expectation of the good news that comes later in the paragraph. To convince the reader, the author shares his conviction through the use of the present tense (in bold in the example).

*In conclusion, our modified gradient vector flow model has not yet provided definitive evidence for or against the use of active contour models in 3D brain image segmentation. However, it **confirms** that polar coordinates, as suggested by Smith et al [4], **are** better than Cartesian coordinates to represent regions with gaps and thin concave boundaries. In addition, we have now removed the need for a priori information on the region being modeled without affecting model performance.*

The findings are inconclusive, but they reveal that (1) an undesirable constraint has been removed; and (2) for a particularly complex type of contour, another coordinate representation scheme is confirmed to be more efficient. Even partial achievements are important to the scientific community when they validate or invalidate other people's theories and observations, and when they establish the benefit of a method against other methods for a particular type of experiment. Science explores, step-by-step, a labyrinth of many dimensions. Marking a dead end before turning back is necessary, especially when much energy has been spent exploring that path.

If the findings are conclusive enough, why wait until all the possible paths have been explored before submitting a paper. Mention what you intend to do next to discourage potential competitors, or to encourage others to collaborate with you.

The 25% improvement in re-ranking the top 10 documents by using words adjacent to the query keywords found in the top five documents demonstrates the validity of our assumption. We anticipate that the high frequency but non-query keywords found in the top five documents may also improve the re-ranking and plan to include such keywords in future research.

In the previous example, the writer stated precisely his future research plan to establish the anteriority of his idea and protect future research.

Presenting in conclusions any limitation that has put a lower ceiling on your high hopes is a perilous exercise. But the ceiling is not permanent — at least, that is what you want to convey to the reader. You know that tackling these limitations is definitely worthy of future research. Relaxing one of your strong assumptions or finding a way to bypass a limitation may enable others to solve their problems. Taking the time to state assumptions and limitations is not only good scientific practice, it is also a way to promote science and your name in science. But how does one constructively present these opportunities in a conclusion? The next example adapted from an IEEE paper starts with a sentence that sends a chill down one's spine.

> *Finally, we summarize the limitations of our optimizing algorithm and offer our future research plan.*

There are so many limitations that the authors find it necessary to summarize them. Two of the authors of the paper were senior fellows, and I suspected they knew how to remain positive in the face of adversity. Indeed, they knew. Here is the first item on their list:

> • *Parameter tweaking. As discussed in section 4.2, the value of alpha is obtained without difficulty, but a satisfactory gamma value is obtained only after experimenting on the data set. We have given the reader pointers to speed up the determination of gamma in this paper. We plan to investigate a heuristic method that allows direct determination of all parameters. In this respect, we believe that Boltzmann simulated annealing will be an effective method.*
>
> • *...*

The parameter tweaking limitation is minimized in two ways: (1) by emphasizing that a method has been given to speed up the labor-intensive part of the algorithm, and (2) by showing confidence that a solution is at hand to bypass this limitation.

Counterexamples

When it comes to conclusions, be conservative and exercise restraint. Do not destroy your good work with sentences like these:

> *In the future, we would like to validate the clustering results not only from the promoter binding site analysis, but also incorporate more information such as the protein-protein interactions, pathway integration, etc, in order to have more convincing and accurate results.*

As a reader, how did you view the achievements? Did you feel that the author was pleased with his contribution?

Here is a familiar sentence written in a humble, self-effacing way, too low key to encourage the reader.

Our method has been used to determine the best terminal group for one metal-molecule junction only, although in principle, it can be applied to other couplings.

Would you trust the conclusions of a paper that ends with the following sentence?

In the future, we intend to experiment our approach using larger data sets.

Does it mean that the current method relies on data sets which the authors think too small?

The next sentence seems fine... if only the writer had not used 'we believe.'

"Although these protocols will continue to change, we believe they represent a reliable starting point for those beginning biochip experimentation."

The positive contribution is placed in the main clause at the end of the sentence, but some readers perceived this sentence as slightly negative. Read this sentence again and skip 'we believe.' You may find the protocols more appealing. The facts appear to speak for themselves, without the need for beliefs to influence the decision of the reader.

In the next sentence, both main and subordinate clauses contain positive facts. Since the main clause contains information about the future, the future should appear appealing. But this is not quite the case:

Although the model is capable of handling important contagious diseases, new rules for more complex vectors of contagion are under construction.

Both subordinate and main clause establish positive facts, yet the overall perception is not always positive. Why? The readers are confused. Ordinarily, if the *although* clause contains a positive argument, readers expect the main clause to negate or neutralize the value of that argument. In this case, the main clause also contains a positive argument. As a result, the overall impression is mixed.

Purpose and Qualities of Conclusions

Purpose of the conclusions for the reader

- They bring better closure to what the introduction announces, by contrasting pre-contribution to post-contribution. What was unproven, unverified, unexplained, unknown, partial, or limited is now proven, verified, explained, known, complete, or general.
- They allow readers to understand the contribution better and in greater detail than in the abstract to evaluate its usefulness.
- They tell the readers that follow-on papers may be expected from the same author.

Purpose of the conclusions for the writer

- They place a particular emphasis on what the contribution allows others to do, immediately or potentially.
- They propose new research directions to prevent duplication of effort, to encourage collaboration, to point to unexplored fields, or to claim anteriority for novel ideas.

Qualities of conclusions

POSITIVELY CHARGED. They maintain the excitement created in the introduction.
PREDICTABLE content. There are no surprises. Everything has been stated or hinted to in other parts of the paper.
CONCISE. Focus on benefits to readers.
Close the door. Open new doors.
COHERENT with title, discussion and introduction. They are a part of one same story.

 Examine your conclusions. How positively charged are they? How consistent are they with the claims you made in the abstract and introduction? Do they "open new doors"?

Conclusions Q&A

Q: If the research stopped, should we still mention something in the future work part of the conclusion?

A: The worst possible thing one could do is to invent future work to have something to write about. You may, however, mention open research issues. The reader will understand that your work has stopped and that you do not intend to pursue these issues, thus leaving them "up for grabs" as they say.

Q: To me, listing the many benefits of my research to the reader feels like a list of sentences starting with 'in addition,' 'moreover,' 'furthermore'... Is there a better way?

A: When you mention these benefits, you may want to give the reader a time frame: the ripe hanging fruit ready for immediate picking without much else to do, the green fruit that will require extra research to come to maturity (your next paper maybe), the flowers pregnant with the promise of a fruit — (a longer term research effort by your group), and the flowers awaiting pollination from other bees (not your research group).

Conclusion Metrics (if you have a conclusion)

✓(+) The conclusion is positively charged.

✓(+) The conclusion is significantly different from the abstract.

✓(+) The conclusion is slightly longer than the abstract.

✓(+) The conclusion does not claim new benefits or findings not already presented before.

✓(+) The conclusion encourages the reader to benefit from the contribution or to further the work.

✓(–) The conclusion presents limitations as a drawback instead of an opportunity to improve.

✓(–) The conclusion does not differ much from the abstract.

✓(–) The conclusion simply restates the results, and if there is an impact statement, it is a plain restatement from the abstract, without elaboration.

AND NOW FOR THE BONUS POINTS:

✓(+++) The reader is able to reconstruct the title from the conclusions.

Chapter 20

Additional Resources for the Avid Learner

First, we would like to congratulate you. If you are reading this chapter, it is because you have decided to further understand the craft of scientific writing by venturing online. We have spent many years exploring the rich humus of the Internet garden and collected many sites worthy of your attention. Do not be surprised if some of the resources seem a little dated — while the engineering and biological sciences advance in leaps and bounds, the science of writing is decidedly more turtle-like. Yet it contains a richness of understanding that benefits modern readers and writers, as it no doubt will for future authors as well. All URLs have been tested at the time of publication. They may change over time but you will always find an up-to-date list on the book site: https://www.scientific-writing.com/the-bonus-page.

Once you have been published, and invited to present your paper at a conference, you will need a new set of skills. Our scientific presentation blog (scientific-presentations.com) will be of great use to you. So will the book we wrote "**When the Scientist Presents** — An Audio and Video Guide to Science Talks". You will find its page on the publisher's site: https://www.worldscientific.com/worldscibooks/10.1142/7198

After you have published a few papers, you may wish to capitalize on them and start to write grants, and here also we can be of help. We have written the book "**The Grant Writing and Crowdfunding Guide for Young Investigators in Science**" which you will find on the publisher's site: https://www.worldscientific.com/worldscibooks/10.1142/10526. Further help on grant writing is also on our blog (www.thesciencegrant.com). Finally, earlier in this book we've made references to (add even quoted from) Jean-Luc's latest book;

a book on writing techniques specifically ccentered on the reader. That book, "THINK READER" goes further into some of the chapters presented in this book. It can be accessed here: https://www. amazon.com/THINK-READER-Writing-Reader-based-techniques/ dp/173389750X. For all matters related to the training we conduct worldwide, you can also reach us at www.scientificreach.com

Epilogue: Your Future Work

Our work ends here, and yours starts now. Writing a book is not easy. Sometimes, only after rewriting and rereading a chapter for the nth time does its structure finally appear. Sometimes, the structure of a chapter is in place even before its contents, and the hard work consists in finding examples and metaphors to make things clear. But one thing is constant: the more you rewrite, the clearer your paper becomes. The memorable words of Marc H. Raibert, President of Boston Dynamics, ex-head of the Leg lab at CMU and MIT still ring in my ears: 'Good writing is bad writing that was rewritten'. How true!

Writing is hard. To avoid making it harder than it already is, start writing your paper as soon as you can. It will be less painful, and even pleasant at times. At the beginning, write shorter papers (e.g. extended abstracts or letters to journals). You can write more of them, and chances are, a few will be accepted. On the way, a few good reviewers will encourage you and pinpoint the shortcomings in your writing, while a few good readers will tell you where you lack clarity.

Each chapter in this book contains exercises that involve readers. Value your reader friends. The time they spend reading your paper is their gift to you. Accept their remarks without reservation and with a grateful heart. Thank them for their help and return the favor (being French, we recommend giving them a bottle of red Bordeaux for their services, but feel free to offer other vintages). Do not take negative remarks personally; instead, consider them as golden opportunities to improve your writing. It is pointless to try to justify yourself because, in the end, **the reader is always right**, and in particular, the reviewers. Just take note of their remarks and questions, and work to remove whatever caused them to stumble. Do not argue.

Let your introduction convey research that is exciting, and let that excitement become your readers' excitement as they look forward to the future your research has opened. Show the world that scientific papers are interesting to read. Create expectations, drive reading forward, sustain the attention, and decrease the demands on your readers' memory. To make reading as smooth as silk, iron out the quirks in your drafts with the steam of your efforts.

Let us end with a parting gift. Once upon a time, in a country where the swan is the national bird, rich in lakes and birch trees, a talented researcher who had attended our class decided to convince some Master's degree candidates in IT that this book contained principles that could be coded inside a free Java application. That application, fittingly for the country, was called SWAN — the "Scientific Writing AssistaNt". Ok, the acronym isn't perfect, but we weren't going to settle on SWA when the the alternative was just so great.

Today, in 2021, 11 years after it became available on the Joensuu university website, amazingly, SWAN still works on Macs, PCs, and UNIX computers. It covers the metrics you find at the end of the chapters in part 2 of this book. Although the interface may look a bit dated, this little app has proven itself fairly resilient to the passage of time. We still use SWAN in the writing classes we, at Scientific Reach, conduct worldwide. SWAN's user interface is not perfect, but its website (http://cs.joensuu.fi/swan/index.html) provides links to help videos on how to use it. A 64-bit version of Java must be installed on your computer for it to run.

Although SWAN is no longer supported, you can contact us through our scientific reach website and, time allowing, we may be able to help you.

May you have a long and prosperous career in research, and may the fun of writing be with you!

Printed in the United States
by Baker & Taylor Publisher Services